MW01224592

Open Range

The Oklahoma Western Biographies
Richard W. Etulain, General Editor

Open Range

The Life of Agnes Morley Cleaveland

Darlis A. Miller

UNIVERSITY OF OKLAHOMA PRESS : NORMAN

Also by Darlis A. Miller

Captain Jack Crawford: Buckskin Poet, Scout, and Showman (Albuquerque, 1993)
Above a Common Soldier: Frank and Mary Clarke in the American West and Civil War from Their Letters, 1847–1872 (Albuquerque, 1997)
Mary Hallock Foote: Author-Illustrator of the American West (Norman, 2002)
Matilda Coxe Stevenson: Pioneering Anthropologist (Norman, 2007)

Library of Congress Cataloging-in-Publication Data

Miller, Darlis A.
Open range : the life of Agnes Morley Cleaveland / Darlis A. Miller.
p. cm. — (Oklahoma Western biographies ; v. 26)
Includes bibliographical references and index.
ISBN 978-0-8061-4117-6 (hardcover : alk. paper)
1. Cleaveland, Agnes Morley, 1874-1958. 2. Women pioneers—New Mexico—Biography. 3. Women ranchers—New Mexico—Biography. 4. Frontier and pioneer life—New Mexico. 5. Ranch life—New Mexico—History. 6. New Mexico—Biography. 7. New Mexico—Social life and customs. I. Title.
F801.C62M55 2010
978.9'04092—dc22
[B]

2009041296

Open Range: The Life of Agnes Morley Cleaveland is Volume 26 in The Oklahoma Western Biographies.

The paper in this book meets the guidelines for permanence and durability of the Committee on Production Guidelines for Book Longevity of the Council on Library Resources, Inc. ∞

1 2 3 4 5 6 7 8 9 10

To the Memory of Myra Ellen Jenkins

Contents

Illustrations

Series Editor's Preface

STORIES of heroes and heroines have intrigued many generations of listeners and readers. Americans, like people everywhere, have been captivated by the lives of military, political, and religious figures and intrepid explorers, pioneers, and rebels. The Oklahoma Western Biographies endeavor to build on this fascination with biography and to link it with two other abiding interests of Americans: the frontier and American West. Although volumes in the series carry no notes, they are prepared by leading scholars, are soundly researched, and include a discussion of sources used. Each volume is a lively synthesis based on thorough examination of pertinent primary and secondary sources.

Above all, the Oklahoma Western Biographies aim at two goals: to provide readable life stories of significant westerners and to show how their lives illustrate a notable topic, an influential movement, or a series of important events in the history and cultures of the American West.

Darlis Miller's smoothly written and revealing biography of Agnes Morley Cleaveland (1874–1958) more than fulfills the goals of volumes in the Oklahoma series. Miller's impressive life story of this vivacious western ranch woman and successful writer will appeal to a full gamut of readers.

Cleaveland often lived life in overdrive. She moved residences often and enthusiastically involved herself in myriad activities. Miller provides inviting glimpses of Agnes's social life, her roles as reformer and activist, and her participation in Christian Science gatherings. As this biography makes clear, Agnes Morley Cleaveland was her own woman—whether within the male-dominated ranch life of New Mexico, the social gatherings of California elite, or clusters of writers and editors.

Miller rightly emphasizes Cleaveland's authorship of her major book *No Life for a Lady.* This well-received book, western folklorist and bibliophile J. Frank Dobie asserted, was "not only the best book about frontier life on the range ever written by a woman, but one of the best books concerning range lands and range people written by anybody." Miller also provides revealing discussions of Cleaveland's other writings and her ongoing struggles to repeat her earlier successes as author.

Independent of mind and action, Cleaveland often lived apart from members of her family, including her husband and children. Sometimes she holed up in Berkeley, California, and immersed herself in a frenetic social life. More often, she retreated to her New Mexico ranch to enjoy the invigorating yet restful sights and sounds of her more remote and arid eden.

In all, Darlis Miller captures the multisided complexity of Agnes Morley Cleaveland and provides an unusually captivating story of a charismatic western ranch woman and social activist. In this marked achievement the author accomplishes the major goal of the Oklahoma Western Biographies series: to treat a notable western person whose life illuminates important scenes of western history.

Richard Etulain
Professor Emeritus of History
University of New Mexico

Preface and Acknowledgments

AGNES Morley Cleaveland (1874–1958) found lasting fame after publishing her memoirs, *No Life for a Lady*, in 1941. The story of her adventures growing up on a cattle ranch in west-central New Mexico captivated readers from coast to coast. The book quickly became a best seller and was widely praised by reviewers. Folklorist and western literary critic J. Frank Dobie called *No Life for a Lady* "the best book on range life from a woman's point of view ever published." Today it is considered one of the classics documenting an important phase of our national history. It often is included on required or optional reading lists in college history courses. And readers everywhere still enjoy Cleaveland's wit and sparkling prose that grace nearly every page.

Yet, who is Agnes Morley Cleaveland? What did she do besides write this book describing her life on the ranch? During her lifetime, Cleaveland's activities often were reported in local newspapers, both in New Mexico and in California, where she spent much of her adult life. By and large, since the book's publication, she has received little scholarly attention.

My husband and I bought property near Datil, New Mexico—Agnes Morley Cleaveland's former stomping grounds—as I was nearing retirement as a faculty member at New Mexico State University. Later, after I had finished another writing project, it seemed only natural that I investigate the Agnes Morley Cleaveland Papers located in the Rio Grande Historical Collections at my university. At first, I was disappointed that the papers seemed to cover only limited aspects of her life. So, in my mind, I settled for concentrating on Agnes's writing career, first as a short story writer, then as author of *No Life for a Lady* and *Satan's Paradise* (1952).

I hadn't counted on Dick Etulain, distinguished literary scholar, convincing me that Cleaveland deserved a biography–that I should take another look at her papers to see if they would support a slender chronicle of her life. He was right, of course. Moreover, I uncovered additional sources, which in conjunction with those at NMSU, have enabled me to flesh out her story and create a biography that I hope will interest readers who have ridden with Agnes through the pages of *No Life for a Lady*.

The first decade of Cleaveland's life was spent in northern New Mexico, where her father, William Raymond Morley, managed the legendary Maxwell Land Grant and Railway Company. He then gained a modicum of fame, when, as an engineer with the Atchison, Topeka, and Santa Fe Railway, he battled to control railroad routes over Raton Pass and through the Royal Gorge. After his death in Mexico following a gun accident, Agnes moved with her mother, brother, sister, and stepfather to a ranch in the Datil Mountains. In her teens, she was sent away to school, eventually graduating from Stanford University in 1900.

Her life from then on was divided between California, where she and her husband, Newton Cleaveland, maintained a home in Berkeley for many years, and the Datil Mountains of what is now Catron County, New Mexico, where Agnes spent her last days on earth. Along the way, she raised one son and three daughters, published scores of short stories, and in the 1930s became a well-known clubwoman and political activist in California. After her husband died in 1944, she moved permanently to her property at Jack Howard Flat near the town of Datil.

In her short stories and memoirs, Cleaveland wrote about the ranching West from firsthand knowledge. She documented both the incredible hardships of ranch life, as well as its many satisfactions. Moreover, women are central to her narratives. For the most part, she depicts them as capable cowhands, independent and self-reliant. In *No Life for a Lady*, she gave an accurate account of her own life as a ranchwoman—one who loved to ride horses and do ranch work. Cleaveland's life away from the ranch illuminates the dynamics of women's social and political networking in the early twentieth century. This part of her story also sheds light on her political conservatism and her friendship with former First Lady Lou Henry Hoover.

I am indebted to several people who helped me complete this biography of Agnes Morley Cleaveland. First, I thank Richard W. Etulain for his strong support and wise editorial comments. The book has benefitted greatly from his careful reading of the first draft and his suggestions for expansion and changes to the text. I also want to thank two of Agnes Morley Cleaveland's granddaughters, Mary Ann Montague and Helen Cleaveland, and Agnes's great-niece, Nancy Reed Miller, for sharing information about their grandmother and great-aunt. And special thanks to Helen Cleaveland and Kelly Gatlin (grandson of Dan Gatlin of the grizzly bear hunts) for introducing me to Jack Howard Flat. Thanks also to Gary Tietjen for allowing me to use his remembrances of Agnes, and to David Caffey and David Adams for sharing documents from their research files about the Morleys.

Jo Tice Bloom and Joan Jensen read all or parts of the manuscript, for which I am most grateful. Also, I thank Sandra Schackel and Sherry L. Smith, the readers for the University of Oklahoma Press, who offered such thoughtful commentary on my manuscript. The following archivists, curators, and club officers provided copies of valuable documents (or gave information): Matthew Schaefer, Herbert Hoover Presidential Library; Jenny Thompson, Alpha Phi Fraternity; David Sawle, California Writers' Club; Sandra Stelts, Pennsylvania State University Libraries; Peter Blodgett, The Huntington Library; Susan Berry, Silver City Museum; and Nancy Brown-Martinez, University of New Mexico Libraries.

Much of the research for this book was conducted at the New Mexico State University Library, especially in the Rio Grande Historical Collections. I owe a huge debt to the following people who filled my requests promptly and cheerfully: Stephen J. Hussman, department head, Archives and Special Collections; Martha Shipman Andrews, archivist, Hobson-Huntsinger University Archives; Leonard Silverman; Dean Wilkey; and Jivonna Stewart, head of interlibrary loan, and her staff.

And, finally, I would like to thank my sister-in-law, Sylvia Walker, for (once again) making research trips so enjoyable, this time providing transportation to Stanford University and to Berkeley, California (where we located Agnes's residence on Cedar Street), as well as providing good conversation and delightful happy hours. And as always, I thank my husband, August Miller, for listening to tales about Agnes, for suggesting that we purchase property near Datil, and for making our trips there everything they should be amidst the ponderosa and piñon pine trees (and the beautiful dark blue day flowers).

Open Range

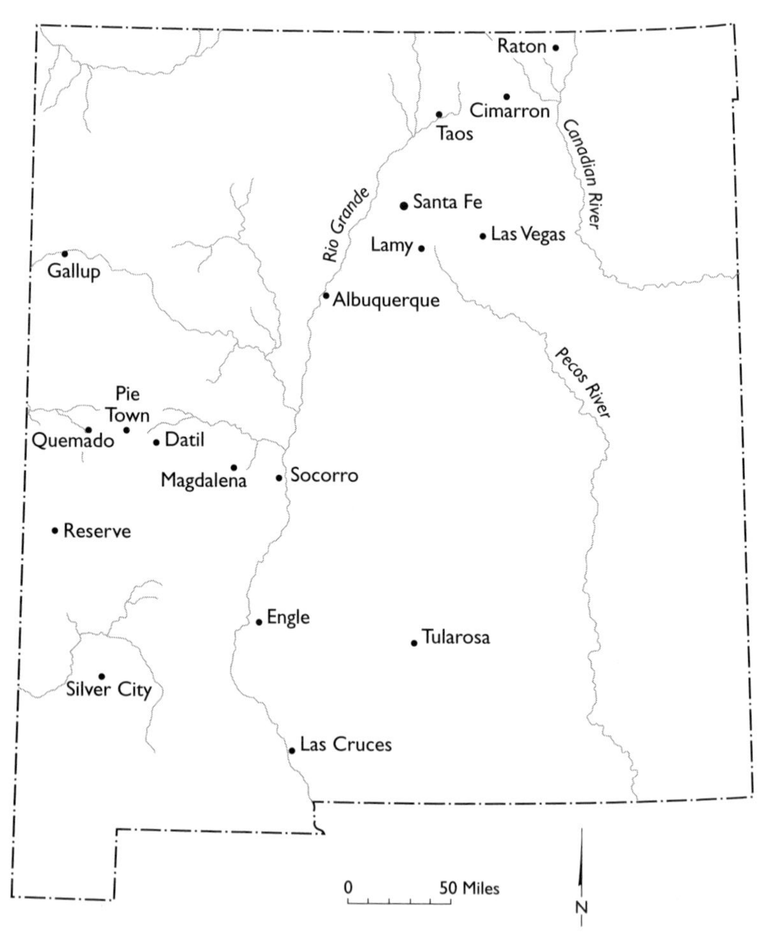

Agnes Morley Cleaveland's New Mexico

CHAPTER 1

A Turbulent Beginning

AGNES Morley was born in Cimarron, New Mexico, on the night that Clay Allison shot up the town. Cimarron, a town of a few hundred residents and headquarters for the legendary Maxwell Land Grant and Railway Company, lived up to its name, which in Spanish means "wild" or "untamed." On the day of Agnes's birth, June 26, 1874, her father, William Raymond (Ray) Morley, then manager of the Maxwell company, had gone into hiding after learning that Allison was coming to town. Given Allison's reputation as a gunfighter, one who held a grudge against Morley, the manager's decision was no doubt prudent. What Morley had not anticipated, however, was the premature arrival of his first-born. Or, as Agnes later described that night in her memoirs, "My father, expecting Clay but not me, had retired to an already prepared hideout, from which he emerged to find Clay gone and the three-odd pounds of my militant self awaiting him."

Agnes's mother, the refined Ada McPherson Morley, had reason to doubt the wisdom of raising a family in Cimarron, especially after four-year-old Agnes witnessed a killing while sitting on the family's front doorstep. Nonetheless, her father, fired with ambition and determined to make his fortune in the West, convinced his young wife to stay on.

The couple had met at the state university in Iowa City in the late 1860s. They came from different backgrounds, but their deep feelings for one another offset such disparities. Born in Massachusetts in 1846 but soon orphaned, Morley grew up on an uncle's farm in Iowa. At age seventeen, he enlisted in the Iowa Volunteers during the Civil War and subsequently marched with Sherman's army to the sea. At the university he took engineering courses but left after two years for lack of funds. In contrast, Ada McPherson, born in 1852, grew up in an Iowa family of privilege and wealth. Her father, Marcus L. McPherson, was a distinguished lawyer and politician, and her mother, Mary Tibbles McPherson, a woman of refined tastes and spiritual leanings. In her youth, Ada became an accomplished pianist, and, after she won a prize in a music festival, her father gave her a Steinway piano, which she took with her when she migrated to the West.

After leaving the university in 1869, Ray Morley carried on a three-year courtship with Ada McPherson, mainly through correspondence. In June of that year, writing from Sioux City, Iowa, where he held a surveying job, he professed his undying love in a long letter to his "darling." He soon headed west, however, and found work with the Kansas Pacific Railway, then laying rails in Colorado and Wyoming. His energy and competence caught the attention of William J. Palmer, managing director of the railroad company. As a consequence, Morley's future—and that of Ada and their children—became entangled in the violence connected with the Maxwell Land Grant and the Colfax County War, and in the dramatic battles to control railroad routes in New Mexico and Colorado.

The story of the Maxwell Grant, a western epic of greed, strife, and political intrigue, centers on Lucien B. Maxwell, who through marriage and a series of purchases came to own the old Beaubien-Miranda Land Grant located in northern New Mexico. By the 1860s, several hundred Hispano settlers lived on this princely domain of immense but as yet unspecified dimensions, many as Maxwell's servants and *peones*. For the most part, Maxwell proved a benevolent landlord, collecting rents of grain, cattle, wool, or sheep without undue discord. As conditions changed, however, and miners swarmed over his property, Maxwell grew weary of managing his estate, and he sold out to a group of European investors in 1870.

Although years of bloodshed and litigation awaited the new owners, they nonetheless expected to profit from their investment. The Maxwell Land Grant and Railway Company was set up to manage local affairs, with William J. Palmer named president of its board of directors. He immediately hired William R. Morley to oversee the company's engineering and surveying work. But after Palmer resigned his position, the company experienced financial difficulties. To bring stability to the enterprise, the board of directors appointed Morley, in November 1872, as the company's manager and executive officer.

By this date Ray Morley and Ada McPherson had become engaged. In accepting this new position, Ray gained the wherewithal to establish a home worthy of his fiancée: a salary of $4,500 per year, lodging in the company's "Cimarron House" (the old Maxwell mansion), and the use of horses and a buggy. He and Ada were married in Iowa in January 1873. On their wedding journey to Cimarron, located in Colfax County in the northeastern corner of New Mexico, they traveled in a stagecoach over the Santa Fe Trail, stopping for a time at Richens L. (Uncle Dick) Wootton's establishment on the crest of Raton Pass. Here Wootton had built an adobe inn and collected fees for the use of his toll road over the pass. As a young bride on her first trip to New Mexico, Ada Morley had no premonition of the heartache that lay ahead for her and her family.

Still, her daughter Agnes would long retain fond memories of the Maxwell mansion, where she was born and raised until the age of five. The two-storied adobe house formed one entire side of Cimarron's plaza. The furnishings were elegant: massive beds and chairs, plush carpets, heavy draperies, large gilt-framed oil paintings, fine china, silver, and crystal, two dining rooms, and four large pianos—two upstairs and two downstairs. Her "most vivid recollection," however, was of the two stuffed Bengal tigers at the foot "of the massive staircase in the entrance hall." "Many a time have I mounted one of those jungle beasts and galloped away to adventure," she later recalled.

Across the street from the Maxwell mansion stood the St. James Hotel, owned and operated by Henry Lambert, President Abraham Lincoln's erstwhile chef. In addition to tending bar, overseeing meals, and welcoming guests, Lambert ran the only livery stable in town. Although townspeople regarded the St. James as a "respectable establishment," many a gunfight took place in its saloon—surely a trial for the young mother, Ada Morley. When guns blazed in the barroom, some patrons found refuge in the Morley residence.

As executive officer of the Maxwell company, Ray Morley inherited a "morass" of financial and legal problems. He also inherited the company's newspaper, the *Cimarron News* (later the *Cimarron News and Press*). To help handle legal problems and to run the newspaper, Morley persuaded his friend and former university classmate Frank Springer to relocate in Cimarron. Among Morley's major headaches were the many settlers who challenged the company's title to the grant and refused to pay rent for or buy the land they occupied. Although Morley and Springer sympathized with the settlers because they faced eviction, they nonetheless agreed that the company held legal title. Both men also came to believe that the Santa Fe Ring—a loosely bound collection of attorneys, politicians, and businessmen who attempted to control territorial politics for their own personal gain—exerted too much influence over company affairs.

The conflagration that erupted into the Colfax County War was ignited by the assassination of the Reverend Franklin J. Tolby on September 14, 1875, as he was returning to Cimarron from the mining community of Elizabethtown. A close friend of the Morley family, Tolby had vowed to "clean up" Colfax County and to break the power of the Santa Fe Ring. Tradition holds that Tolby and Morley had written an anonymous series of articles for the *New York Sun* that excoriated the Santa Fe Ring.

After Tolby's death, violence escalated. Oscar P. McMains, a part-time Methodist preacher, vowed to track down Tolby's killers. In a string of tragic events, vigilantes hanged the suspected killer, Cruz Vega, and killed two others, one a relative of Vega and the other thought to be connected to Tolby's death. In the midst of these tumultuous times, Ray Morley left for Las Vegas, New Mexico, on business, and Ada and little Agnes found refuge at the ranch of Manley and Theresa Chase, the Morleys' good friends.

Tensions in Cimarron continued to mount. Residents directed some of their hostility against the *Cimarron News and Press*, which, under editor Will Dawson, now sided with the Santa Fe Ring. Ray Morley and Frank Springer, although no longer in control of editorial policy, had become joint owners of the newspaper. So when a mob broke into the *News and Press* office and demolished the press on the night of January 19, 1876, it was a financial loss for both men.

In her reminiscences, Agnes Morley Cleaveland recalled that it was Clay Allison and his henchmen who destroyed the press and dumped it into the river. The next morning, however, Allison made amends after he encountered Agnes's mother, now seven months pregnant, at the site of the wrecked building. According to Agnes, Ada lambasted Clay for his malicious deed. Thereafter, he stuck a roll of greenbacks into Ada's hand, telling her to go buy another printing press. "I don't fight women," he reportedly said.

In the spring of 1876, a grand jury indicted Reverend McMains (and fourteen others) for the murder of Cruz Vega. Thereupon, McMains was arrested and placed in the Cimarron jail to await trial. His friends often visited him there to help relieve his boredom. On one occasion, Ray Morley brought Agnes along to receive a little bow and arrow that McMains had made for her. On other occasions, McMains was let out of jail to take meals with the Morleys and other supporters. When the case finally came to trial in August 1877, McMains was convicted of murder in the "fifth degree" (meaning McMains's negligence resulted in the killing of Vega). The verdict was later quashed.

In the year preceding McMain's trial, the Morley family had undergone several changes. Ada gave birth to a son, William Raymond Morley, Jr., on March 17, 1876, and later that spring she traveled east with the baby and two-year-old Agnes to visit her family. In the summer, Ray left the employ of the Maxwell company to work for the Denver and Rio Grande Railway

(D&RG), then constructing lines in southern Colorado. A year later, however, he switched his loyalties to the Atchison, Topeka, and Santa Fe Railway (AT&SF), which had extended its rails to Pueblo, Colorado. As the Santa Fe's locating engineer, Morley became a key player in its fight with the D&RG to control two desirable railroad routes, one over Raton Pass providing access to the West Coast and the other through the Royal Gorge (Grand Canyon of the Arkansas) leading to the rich mining district of Leadville, Colorado.

While Ray was in the field, often for months at a time, Ada and the children continued to reside in Cimarron. The senior Morleys kept up a steady stream of correspondence when apart, writing of their activities and expressing affection for one another. On September 26, 1877, for example, Ray wrote to "My Dear One" from his camp near Florrissant, Colorado: "I have received nine letters from you since writing but you will not blame me I know for I have been almost constantly either in the saddle or on the line from morning at sunrise until dark." He was "rushing things as fast as possible" so that he could return to his wife and babies. Four days later he wrote again, saying that his location work was going well and that he would be in Raton Pass by mid-October—and then home. "I thank God for the faith I have that I shall meet this same good noble pure wife that I left in July," he concluded.

Legend has it that Morley spent weeks scouting the Raton area disguised as a sheepherder so that rival D&RG surveyors would not suspect his real mission. He wandered over the hilltops with Uncle Dick Wootton's sheep and stopped at Uncle Dick's inn for provisions. On December 19, he notified company officials that his survey was finished and that Wootton would sell the right-of-way to his toll road for a fair price.

Ray probably did rush home after finishing his work in Raton Pass—to be with Ada when she gave birth, on January 9, 1878, to their third child, Ada Loraine (Lora) Morley. Still, in February, when William B. Strong, general manager of the AT&SF, ordered Albert A. Robinson, his chief engineer, to occupy Raton Pass, Ray went with him.

To keep from arousing the suspicion of rival railroaders, Robinson and Morley left El Moro, Colorado, a D&RG townsite, in the dead of night in a hired buckboard. Once they reached Uncle Dick's place some twenty miles away, they enlisted a makeshift crew of cowboys and freighters camped nearby to start work on a railroad bed. A half hour after sunrise the next morning, February 27, a D&RG work crew arrived to find their rivals already in control of the pass. Heated words were exchanged, but Robinson and Morley stood firm. For the rest of the year, Robinson's crews labored to build a road from the AT&SF line at La Junta, Colorado, to Raton. When the first rail was laid across the Colorado border into New Mexico, on December 6, 1878, Ray Morley was on hand to take part in driving the ceremonial spikes.

Meanwhile, six weeks following Morley and Robinson's race to control Raton Pass, Morley made a second historic ride to claim the right-of-way through Royal Gorge for the AT&SF. At stake was access to the mining camp of Leadville, high in the Rocky Mountains, where a silver boom had begun in 1877. Both the AT&SF and the D&RG wanted to serve this area of great potential wealth. The only feasible route was through the Royal Gorge, a spectacular crevasse whose walls in some places rise 3,000 feet above the river below.

The D&RG seemed to have the advantage, as its lines extended to within a mile of Cañon City at the mouth of the gorge. On April 16, 1878, company officials ordered a construction crew to move by train from El Moro, via Pueblo, to the end of the track at Cañon City and then begin building into Royal Gorge. Learning of these plans, AT&SF officers told Ray Morley to get to Cañon City first. The race was on. Morley, who had been in La Junta, rushed to Pueblo where he hoped to charter a D&RG train to reach his goal. Rebuffed by a D&RG agent who refused his request, Ray saddled his own big black horse, King William, and dashed off into the night.

Morley and King William made the forty-mile ride in record time and reached Cañon City ahead of the D&RG work train. He then roused the townspeople to action. They despised the

Rio Grande company and turned out en masse to help Morley. As the *Colorado Chieftain* of Pueblo reported: "Morley had no men with him, but the people of Cañon sympathized with him, and it was but the work of a few moments to get every available man and boy in the city to shoulder a shovel, gun or pick. All the available teams were collected and the party, headed by Morley, who had made a regular Phil Sheridan ride of it, rushed up to the mouth of the cañon and commenced grading on the line surveyed last fall through the cañon."

A half hour after Morley's men entered the cañon, a D&RG crew arrived and commenced work a few yards away. Each side quickly brought in gunmen to solidify its claim to the right of way. "Tensions increased, tempers flared," and a real war seemed possible. In her memoirs, Agnes Morley Cleaveland succinctly wrapped up this episode in her family's history: "Months of litigation and some physical violence, known as the 'Grand Cañon War,' ensued. The dispute was finally settled by the purchase of the Santa Fe's rights by the Denver and Rio Grande." She then inserted the one memory she retained of the entire affair: the grand moment when her father led her to the brink of the gorge and allowed her to peer down into its depths.

In mid-1879 the family moved to Las Vegas, where Ray, as the Santa Fe's newly appointed resident engineer, had his headquarters. But Morley seemed always on the move, and by year's end, he had begun work on another major engineering project–to locate and build a railroad from Guaymas, Mexico, a small fishing village on the Gulf of California, to Nogales, Arizona. With connecting rails extending into New Mexico, company officials hoped to create a new transcontinental route that would capture a large share of foreign trade funneled through Guaymas.

For the next three years, Morley spent most of his time in Mexico, joined on occasion by Ada and the children. In early December 1879, he wrote to her from Hermosillo in the Mexican state of Sonora, reporting on the progress of his location work. On New Year's Eve, upon his return to that city from a brief trip to Guaymas, he received word that Ada had given birth to a baby boy, whom they named Frank. Clearly overjoyed by the news, he nonetheless told Ada in a letter written January 1, 1880, that he regretted not having been "with you dear to help and cheer you in your hours of pain and peril." He then speculated on the reactions of the other children to this newest member of the family: "I can hear Agnes discussing this matter in her wise way and Ray's cute remarks, but poor little Lora—how she must have rebelled and moaned when she lost her baby rights to another."

That summer the family spent time on the Pacific Coast before embarking on a sea voyage to Guaymas. The U.S. census for 1880 lists the Morleys as living in Oakland, across the bay from San Francisco. Baby Frank was now six months old, the other children, starting with Agnes, six, four, and two. Also living on the premises was a seventeen-year-old woman from Norway working as a housekeeper. Ada Morley became enraptured with the California countryside. "I have seen all its loveliness of bloom," she later wrote, "and wealth of nature—grandeur of the sea and charming scenery of vale and foothill . . . full many a charm I have enjoyed there."

Ada found little but heartache in sweltering Guaymas, however. "We left lovely California and went to Guaymas," she wrote to a friend, "and do you wonder, I said, I went from heaven to hell." Indeed, tragedy struck soon after the family's arrival, for little Frank died on August 21. Ada Morley lived through her grief, caring for her children and supporting Ray's work, and did not return to California until the following April.

Ada and the children's next trip to Mexico probably took place in October 1881, when they joined Ray in Tucson, Arizona, and then embarked upon a carriage ride of roughly 250 miles south to Hermosillo, where they evidently lived for several months. Morley completed the railroad line from Guaymas to Hermosillo in November of that year and then "pushed construction north toward Nogales." During her stay in Hermosillo, Agnes become best friends with a little playmate named Cuca. Years later, after Agnes published her autobiography, Cuca wrote to the now-famous author: "I still have in my memory many of the happy days we had." She also treasured Agnes's picture, the one Agnes had given to her when she went away.

Evidently, Ada and the youngsters spent the summer of 1882 in New Mexico and then traveled to Nogales in October for the festivities held to celebrate the opening of the Guaymas-Nogales line. Only eight years old at the time, Agnes later recorded her recollections of the day in her memoirs. "The town is in gala array. Flags and bunting are everywhere," she wrote. Her father, "wearing a long frock coat and a high silk hat," made a short speech; her mother, in her "voluminous silk skirts . . . and a little bonnet gay with flowers," struck with "one ladylike blow" the last spike in the Guaymas to Nogales track. More speeches followed, and presentation of gifts, "the spike and a gold watch to my father, a pair of ponderous and ornate gold bracelets to my mother, and, amid much laughter, a pair of small gold filigree earrings to me." There were gifts for Ray and Lora as well.

All too soon, tragedy again befell the family. Following his triumph at Nogales, Ray Morley signed on to build the Mexican Central Railroad, which would connect El Paso, Texas, with Mexico City. On January 3, 1883, however, while riding in a carriage in Mexico, Morley died after a rifle, entangled in the driver's reins, accidently discharged, sending a bullet into his aorta. That explosion, his daughter later avowed, "brought the firmament crashing down upon a little family's head."

CHAPTER 2

Growing Up in White House Canyon

IN an unpublished essay, the sociologist Mary Roberts Coolidge (wife of author Dane Coolidge and friend of the Morley family) described Ada McPherson Morley as "the intellectual and wholly unpractical" wife of William R. Morley. Similarly, Agnes Morley Cleaveland, in *No Life for a Lady*, depicted her mother as being "tragically miscast" as a rancher, a role thrust upon her by the death of her husband. She spent too much time writing letters in support of "causes"—temperance, women's suffrage, animal rights, for example—and too little time overseeing the stock business. Both Coolidge and Cleaveland slighted those character traits—tenacity, resilience, and spunkiness—that allowed Ada Morley to build a new life for herself and her children after that tragic accident in Mexico.

A large crowd turned out for Ray Morley's funeral, held at the new Plaza Hotel in Las Vegas on January 8, 1883. Friends and neighbors "filled the hotel's ballroom and overflowed onto the sidewalks," one observer reported. Among the pallbearers were Ray's good friends, Henry M. Porter, Frank Springer, and Dr. J. M. Cunningham. An obituary in the *Las Vegas Gazette* avowed that the deceased left "a comfortable fortune, accumulated during the years of his toil in the new Southwest, which insure [*sic*] his family ease and comfort so far as worldly goods may avail." Unfortunately for Ada and her children, this prophecy did not come to pass.

Immediately after the funeral, Ada gathered her children and took refuge at the ranch of Manley and Theresa Chase. The Chases were among the largest cattle ranchers in northern New Mexico, having arrived in the territory in 1866. Their ranch headquarters on Poñil Creek, a few miles northeast of Cimarron, had often provided a haven for the Morley family prior to Ray's death. Tradition holds that Ray and Manley together organized the Monte Revuelto Cattle Company and had made plans for its development.

Now Ada turned to Manley Chase for help. Upon leaving the Chase ranch, she traveled to Red Oak, Iowa, home of her McPherson relatives, where, on January 30, 1883, she wrote to Manley about business matters. She recently had learned that Ray had taken steps to buy a mine in Sonora. Should she carry out these plans, she asked Manley, or put her money into cattle? "You know where money will grow better than any friend I have on earth," she declared. Most of all, she continued, "I want to carry forward as much as possible all [Ray] began."

By mid-March, Ada had returned to Las Vegas, where she and the children moved into a house on "aristocratic hill." With the arrival of the railroad in 1879, the town experienced a population and commercial explosion. The number of townspeople had increased from 1,730 in 1870 to 4,697 a decade later. Prominent mercantile firms—Gross, Blackwell and Company, Charles Ilfeld and Company, and Romero Mercantile Company—helped establish Las Vegas as one of the leading cities of the territory. This bustling community must have seemed to Ada a suitable place to raise the children and to look after her husband's interests.

A stream of friends soon called upon the grieving widow, although Ada wrote to Theresa Chase that she did not go out at all. She made an exception, however, to hear Bishop George K. Dunlop on "Is Christianity a Failure?" and admitted that "as blow has fallen upon blow I have had cause to question the motives of God himself." Still, her hopes regarding the future of her children were "beginning to grow." She was teaching them at home, where she had "every facility," and they were "very bright scholars."

In the days to follow, Ada continued to rely on Manley Chase for financial advice. On September 14, 1883, she and Manley jointly bought out John B. Dawson's share in the Red River

Cattle Company, one of the seven cattle companies that Chase managed. Ada also acquired shares in the Gila, Monte Revuelto, and Luera cattle companies, all under Chase's management. In a letter to Chase, dated January 6, 1884, she expressed her hopes for the future. She had just returned from a 1300-mile round trip to view the spot where Ray had been killed. Now she resolved to live for her children. "I want to make the most of my means for their sakes," she wrote. "I want to have enough to give the girls at least—an independence so they can choose a life partner from a heart choice and not marry for money as so many do." She was determined "to save and invest," following Manley's advice.

The first glimpse we have of young Agnes Morley during the time she lived in Las Vegas appears in the opening pages of her memoirs. Here she described her role in Decoration Day ceremonies, held on May 30, 1884, at the Odd Fellows' Cemetery, where her father was buried. She was one of forty-eight little girls in white dresses that carried American flags and accompanied a corps of Civil War veterans who placed flowers on the graves of fallen comrades. What is significant about this remembrance is that it casts Agnes in a role that she frequently repeated throughout her life—a performer in front of an appreciative audience. Later that same year, in November, Agnes appeared in a program sponsored by the Woman's Christian Temperance Union, in which she recited "The Two Glasses"—one filled with wine, the other with water—before a crowd of two hundred.

By the time of her husband's death, Ada Morley probably had made the acquaintance of Floyd Jarrett, a shareholder in the Gila Cattle Company, who also would supervise its operations in west-central New Mexico. A native of Virginia, he no doubt encountered the lovely widow a number of times at the Chase ranch and began to court her. Evidently, Jarrett charmed little Agnes as well as her mother. About a year after her father's death, Agnes wrote to "Mr. Jarrett my dear friend," and expressed warm feelings toward him. She signed the letter, "Lovingly, Your little friend Agnes."

In her memoirs, Agnes Morley Cleaveland had very little to say about the man who became her stepfather, never revealing his name, even though he had been part of the family for a handful of years. Agnes portrayed Jarrett as "a persuasive and soft-spoken gentleman," whose "bounding optimism was stimulating" but whose "fondness for hard drinking and gambling was unsuspected." Agnes's sister Lora also recalled that her stepfather drank steadily "but was never drunk that we ever saw—he was always the 'perfect gentleman' when full. Never vulgar, profane, unkind or disgusting." His debts for gambling and drinking would add to their mother's financial woes.

The marriage between Ada Morley and Floyd Jarrett took place in mid-November 1884 in Kansas City, an event that several of Ray Morley's friends looked upon with disfavor. Loyal to the core, these friends had collected donations to erect a monument to grace his burial plot in the Odd Fellows' Cemetery. This pillar had arrived in Las Vegas shortly before Ada left to be married. She had been told, however, that no special ceremony was planned for its unveiling because so few donors were in town to attend. Later, she was shocked to learn that a public ceremony had taken place almost simultaneously with her marriage. Because the timing of Ada's second marriage appeared to many townspeople as a "wanton violation of good taste," both she and the editor of the *Las Vegas Daily Optic* published statements in the press to explain how this unfortunate coincidence occurred. With some bitterness, Ada expressed her belief that the organizers had deliberately withheld news of the unveiling because they wished to hold her up "to public ridicule and derision"—all because they objected to her marriage. In closing, she reminded readers that Ray Morley, with "the faultless beauty" of his life and character, would "resent anything intended to work injury to those he loved most."

By the time of their marriage, Floyd and Ada Jarrett likely had decided to purchase a ranch in the open rangeland of western Socorro County, New Mexico. According to Agnes, her stepfather had dreams of becoming "one of the cattle barons of the day" and persuaded her mother to invest most of her available cash in a huge ranch in the Datil Mountains. The Jarretts, in fact, were typical of many entrepreneurs who poured money into the western cattle industry during the 1880s. Newspapers of the day touted the ease with which fortunes could be made: cattle grazed freely on public lands, ranch hands worked for low wages, and the only land that needed to be owned outright was that which controlled sources of water. Certainly, Floyd Jarrett's work with the Gila Cattle Company had provided the opportunity to spy out good prospects.

In *No Life for a Lady*, Agnes wrote that the family moved to their ranch in the Datil Mountains in February 1886. But evidence suggests that she was in error, that the family made its momentous trek a year earlier. On December 10, 1884, Manley Chase wrote to an associate that the Jarretts, who recently had returned to Las Vegas, "intend to move out to a ranch in the Datils, near the Gila." And the *Daily Optic* reported on December 24 that "Mr. Floyd Jarrett and family will remove to the ranch, eighty miles west of Socorro, next week." Ada's letter to Chase, dated February 2, 1885, however, makes clear the date of their departure: "Today we leave for Socorro Co., to reside permanently in all probability—and I want very much to build and purchase cattle and get settled down."

The Morley-Jarrett family traveled by train to Socorro, the county seat, situated on the banks of the Rio Grande about two hundred miles southwest of Las Vegas. With arrival of the railroad in 1880, Socorro was booming; miners, speculators, and businessmen had flocked to the area, boosting the town's population to 4,075 by mid-decade. Among its prominent citizens was German-born Gustav Billing, who in 1883 built a smelter on the outskirts of town to handle lead carbonate ores from his mine, the Kelly, located in the Magdalena Mountains some thirty miles to the west. To tap the wealth of the Magdalena mining district, the AT&SF built a spur (completed in early January 1885) from Socorro to the new town of Magdalena, located three miles from Billing's mine.

The family would have boarded the 3:45 P.M. train in Socorro and then traveled slowly up some steep grades before reaching

Magdalena, where they spent the night. Agnes recalled that her mother asked the hotel keeper to "give us a room that is not directly over the barroom," fearing that bullets fired by boisterous patrons would "come up through the floor." Most likely, they stayed in a combination saloon-hotel, where upstairs rooms were rented to travelers. In the mid-1880s this raw railroad town consisted of only a few houses, stores, and saloons, but in the coming years Agnes would see it develop into a major distribution and shipping center for a vast region to the west.

Bundled up in quilts to ward off the cold, the family left town the next morning in a buckboard drawn by four horses. Their destination was Baldwin's stage station nearly thirty-five miles west at the foot of the Datil Mountains. Agnes's memories of this day are much more sanguine than her sister's. Agnes recalled that "the sun was warm, and the air had that crystalline quality which puts a sparkle in your spirits as well as ozone in your lungs." In contrast, Lora remembered that it was cold, that the quilts "didn't keep us warm." The wagon road led west up an incline and through a stretch of piñon and juniper trees before topping a divide and then cutting straight across the San Augustin Plains, "an irregular ancient lake bed, fifty to seventy miles in cross distances," to the mountains beyond. This was open range country, soon to be dotted with herds of cattle and flocks of sheep. It also was home to large droves of antelope; perhaps as many as four thousand surged into view as the family traveled west.

Once across the plains, they spent the night at Baldwin's, a combination stage station, store, and hotel, run by Levi Baldwin and his wife, Mary. The next morning they set out for the remaining ten miles to the ranch, much of it through Datil Canyon, its steep slopes covered in piñon, juniper, and ponderosa pine trees, its lateral valleys carpeted with clover and grama grass. This strikingly beautiful country seems to have captured Agnes's heart the moment she entered the canyon.

In her memoirs, Agnes vividly described the children's exuberance as they approached their final destination. Eight-year-old Ray jumped onto the seat of the wagon to be the first to catch a glimpse of their new home as they cleared the final rise. "There she is!" he whooped, pointing to a two-room log cabin, situated on the canyon floor at an elevation of 8,300 feet. Not to be outdone by her little brother, ten-year-old Agnes screamed, "That's *my* mountain! . . . And that's my cañon," sweeping her arm in an arc to claim part of the landscape. This would be her country, she later reflected, "for so long as I should live." Agnes's mother, less enthusiastic no doubt about the cabin and new living arrangements, soon had Navajo blankets strung across the larger room, making two rooms into three.

Work began almost immediately on a spacious two-story, ten-room log house. Logs were felled on "the near-by mountain-sides," Agnes recalled, and dragged by ox teams to the canyon bottom. Four expert axe men were brought from Michigan to hew the logs. When completed, the inside walls were plastered "glossy white and hung with oil paintings and family portraits in heavy gilt frames." Deep-piled carpets covered its floors and "lace curtains hung at the high windows." Ada's Steinway dominated one room, and walnut bookcases overflowed with books, many from Judge McPherson's library. Because the outside doors, window frames, and veranda pillars were painted white, locals began calling the place "the White House"—and Datil Canyon today appears on maps as White House Canyon.

A *Socorro Bullion* correspondent, who visited the Jarretts soon after completion of this imposing structure, described it as "one of the most picturesque and pleasant mountain homes I have yet found in the West. Money does not seem to have been spared in purchasing and fitting [it] up." The gracious Mrs. Jarrett had ushered him into the sitting room, adorned with costly ornaments. "It was with reluctance," he told his readers, "that I bid these estimable people farewell."

Agnes later claimed that this "pleasant mountain home" was totally impractical, especially in the winter when temperatures often dipped below zero. It took one man "to haul and chop wood" to burn in the fireplaces "to keep that high-ceilinged, spacious-roomed house even approaching livability. . . . And the minimum of good housekeeping standards took the entire time of one woman," she added. Moreover, the housekeepers her mother employed inevitably departed shortly after they arrived.

Floyd Jarrett evidently spent much of his time scouting the countryside for choice property. Ada admitted in a letter to Manley Chase, written in July 1885, that she had no idea where Mr. Jarrett was–"out west hunting more ranches—springs I guess." His propensity to roam may explain his absence from the ranch when an "Indian scare" electrified the area.

On May 18 messages were telegraphed across the region that Geronimo and about fifty Chiricahua Apaches had fled the San Carlos Reservation in Arizona. This feared Indian fighter led part of the band south into Mexico; the remainder, under Chihuahua, headed east into New Mexico, where they "divided into small parties which raided in widely separated localities." General George Crook, in charge of military operations, soon had soldiers combing southwest New Mexico. Five troops of the Tenth Cavalry moved north of Silver City toward the Datil Range. Local newspapers kept residents apprised of the situation, keeping count of casualties, although the number of civilians killed during this outbreak is not known for certain. On May 26, the *Socorro Bullion* claimed that the "Apache Devils" had killed eight people in the Mogollons northwest of Silver City; a week later it trumpeted: "The blood of no less than sixty recent victims of Apache cruelty and Government stupidity cry to heaven for vengeance."

During this difficult time, a courier dashed up to the Jarrett cabin and shouted, "Geronimo is this side of Quemado!" Even though ten heavily armed men were at work on the new house, Ada Jarrett insisted on taking her children to Baldwin's for protection. A born storyteller, Agnes engagingly described in her memoirs the flight of the family down the wagon-road to Baldwin's. Only two horses were available, forcing them to leave without an armed escort. In Agnes's words:

> Two on a horse is never comfortable, less so when the pace is an unrelieved high trot, which was as fast as we could force our ponies to go. Lora, riding behind mother, valiantly

stifled her moans, and the robust nine-year-old Ray gouged his fingers into my midriff as he clung to me, muttering child profanity under his breath as he bounced up and down on the stiff saddle skirts which projected beyond the cantle.

They reached Baldwin's safely, finding the place crowded with other "terror-stricken families" who had come in from distant ranches. Sleep was almost impossible that night, as worried parents kept watch over children bedded down on the floor or, in the case of the youngest children, on whatever beds were available. With daylight came the welcome news that the Apaches were far to the south, heading for Mexico.

In following months, Crook's troops successfully pushed most of the off-reservation Apaches back to San Carlos. Geronimo remained at large, however, as did a small number of other Chiricahuas who continued to raid for the supplies they needed to keep alive. To protect settlements in west-central New Mexico, the military established posts at Horse Springs, about thirty miles southwest of the Datils, and at Datil Creek near Baldwin's station. On October 7, 1885, Captain Charles J. Dickey, commanding officer at the "Camp on Datil Creek," reported, "The country north, east and south has been thoroughly scouted, every ranchman within 25 miles of my camp notified to look out for Indians."

In November, Apache raids in the Arizona–New Mexico border region again thoroughly alarmed local residents. On the twenty-first, the *Socorro Bullion* proclaimed, "The Apaches during the last raid robbed six ranches, burned two, killed ten people and wounded three dangerously." This same issue announced that Floyd Jarrett had been excused from jury duty, "owing to the presence of hostile Indians in the vicinity of his ranch." Yet this Indian scare in the Datils remained just that—a scare. Captain Dickey reported in early December, "No Indians or any signs discovered," and several weeks later, "There is no Indian news." The military finally withdrew troops from Datil Creek shortly after Geronimo surrendered in September 1886.

Despite threats of Apache raiding parties, Ada and Floyd Jarrett continued to expand their ranch, often on borrowed capital. Shortly after their arrival in the Datils, Ada reported in a letter to Manley Chase that she had been "borrowing for running expenses and building a home . . . we bought a ranch west of us we couldn't possibly let go by an opportunity that wont come twice in any man's life and borrowed again at ruinous interest to pay for cash down."

In November 1885, Ada signed a power of attorney to allow her husband to represent her in business matters. Two months later, he purchased twelve hundred head of cattle from G. N. Gentry of Hamilton, Texas, signing promissary notes to Gentry amounting to more than $10,000, with the cattle as security. Ada made a valiant effort to keep informed of her investments, as attested to in her correspondence with Manley Chase. But she must have suspected early on that granting a power of attorney to her husband had been unwise, for she had it revoked within two years of signing the document.

Floyd Jarrett's rise in the cattle world did not go unnoticed by the local press (although it totally ignored Ada's role in financing his success). On November 7, 1885, for example, the *Socorro Bullion* reported, "Floyd Jarrett is building a $10,000 residence on his ranch in this county." And, after announcing Jarrett's purchase of Gentry's cattle, the paper called Floyd "one of the biggest and best cow men in the territory."

The Morley children, too young to understand fully their mother's financial problems, thrived in this new environment. "Life on a cattle ranch of that day," Agnes later wrote, "was any child's idea of heaven. . . . Nothing we could think up to do mattered much to anybody, certainly not the neighbors. The nearest of ours was several miles away." With adults preoccupied with running a business, the children were free to explore the countryside—often disappearing for hours without causing anyone undue anxiety.

At an early age, Agnes developed into a self-reliant and resourceful youngster, who took special pride in the work she did on the ranch. One of her daily chores was to bring horses from the pasture into the home corral before breakfast. At a time when fences were almost nonexistent, hunting horses on foot required all the patience and stamina she could muster. She performed other errands on horseback. "Put a kid on a horse" was the usual method of sending messages between ranch houses or of delivering supplies to outlying camps. After a weekly mail service was established between Magdalena and Baldwin's (thereafter known as Datil), Agnes and her siblings made the twenty-mile round trip, either alone or in pairs, to collect the mail every Monday—"in rain or shine, heat or cold, daylight or night-time."

The Morley children rendered other valuable services as well. They were expected to be observant, to report what they had seen when riding across country. They learned to recognize characteristics of individual cattle, their brands and markings, and to know where their own cattle were grazing. They learned also to report the presence of a maverick, for finding one of these unmarked calves was "like finding a gold nugget." And they hurried to report a cow that had bogged down in mud, a calamity that might claim the victim if not pulled out in time.

Like other ranch children, the Morleys faced a variety of potential hazards on their cross-country rides, including the grizzly bears, black bears, bobcats, and cougars that still prowled the forested mountainsides. Once, when Ray was delivering a message on horseback, he suddenly came upon a mother grizzly with two cubs. Fortunately, the bear chose not to attack and ambled down the trail with Ray slowly following after her. A short time later, Agnes traveled over the same route—and was chagrined to learn that she had mistaken the grizzly tracks for the footprints of a sheepherder. On another occasion, Agnes met up with a dangerous creature of a different sort—an outlaw badly in need of a fresh mount to outdistance a sheriff. Instead of taking hers, however, the chivalrous bandit allowed her to ride on up the trail and was later gunned down while trying to steal a horse from Baldwin's horse pasture.

Several other times the Morley youngsters demonstrated a courage that belied their age, as, for example, when eleven-year-old Ray discovered early one morning that Indians—possibly Navajos or Pueblos—had stolen his pony. After tracking them for hours on horseback, he rode boldly into their camp and retrieved his property. In his eyes, he had done nothing exceptional; if he had wasted time going for help, he explained, the pony would have been lost for good. At the age of twelve, Agnes also encountered Indians while riding alone on the ranch. She was enthralled by the sight: forty Navajos clad in buckskin riding toward her in single file. Slowly and silently, they rode in a circle around her—fascinated, as it turned out, by her long, light-colored hair. "[It] did not occur to me to be frightened," Agnes reported. "I was as interested in the Indians as they were in me."

The two older Morley children were daredevils and challenged each other to perform acts of bravado that sometimes ended in bodily harm. "Betting the other he was afraid to get on some horse of dubious character was our favorite sport," Agnes recalled. After she accepted such a challenge and her horse reared and flipped over backward (leaving Agnes bruised but otherwise unhurt), no one thought to tell the grown-ups. On another occasion, Agnes accompanied friends on an exploration of a gypsum cave. With candles to light the way, they moved forward cautiously until they reached a room whose only entry was through "a mammoth rat hole," too small, however, for adults to navigate. Agnes volunteered to crawl though on her stomach, with two men holding onto her feet. She carried a few matches in her mouth and an unlit candle in her hand. Once she squeezed through the hole and lit the candle, her hair caught fire. Frantically, she beat at her head with her hands and screamed; the men quickly jerked her back, "torn and scratched . . . badly singed, although fortunately not badly burned."

During these seemingly carefree days, the Morley children resumed their schooling under the watchful eye of their mother, a woman "who believed more than anything else in education and culture." In the spring of 1886, Ada moved to Socorro with the youngsters so they could attend the Socorro Academy, a private school of about twenty pupils of all ages.

The following year, she hired a governess, Mary Avey, to tutor them at the ranch. Years later, Agnes took delight in the letter her former tutor wrote to congratulate the author upon publication of *No Life for a Lady*. "It was my great pleasure to be a guest at your beautiful ranche [*sic*] home one time," Mary Avey declared, "where everything was so new and wonderful to me, every one so kind, surroundings so interesting and restful." She then recalled the time that she and Agnes made a long ride to notify a rancher that his cow had bogged down on Jarrett property and "could not be coaxed . . . to help herself out." The rancher solved the problem by shooting the animal.

In the early years on the ranch, Ada Jarrett encouraged the children to read books in her library, the "classics," which she had brought with her over the Santa Fe Trail. But she kept close watch on what they read. As Agnes later wrote, "[Mother's] wide laxity in what we were permitted to do [on the ranch] did not carry over into what we were permitted to take into our minds by the way of the printed page. Not until I went away from home, I think, did I ever read anything without first asking her permission." She remembered having read *Jane Eyre* and *Vanity Fair* from her mother's library, as well as the works of Washington Irving, James Fenimore Cooper, and William Cullen Bryant. The family also subscribed to innumerable papers and magazines, some of which the children would enjoy reading, including *St. Nicholas*, *Youth's Companion*, and *Chatterbox*.

At age fourteen, Agnes received a "sentence of banishment" when her mother sent her east to attend the Friends Central School in Philadelphia, a school highly recommended by Ada Jarrett's Socorro neighbor and good friend May Miller, wife of the town's chief dentist and a former resident of Philadelphia. Agnes was to board in the household of the Reverend Charles Gordon Ames, pastor of a Philadelphia Unitarian Church. Agnes later remembered the Ames's home as one of "supreme culture," whose visitors included such personages as Edward Everett Hale and Elizabeth Cady Stanton. And, although she described herself as "a self-conscious, wild eyed ranch child," the Ames family (which included two daughters) and the Quaker teachers and students of Friends Central made her feel right at home.

Agnes studied diligently while in school, immersing herself in Latin, world history, composition, and other subjects. An outgoing youngster, she made friends easily and enjoyed telling stories about life on the ranch. As would happen quite often later on, city dwellers wondered if her tales could be true. After reading Agnes's essay, "A Wild Horse Hunt," Annie Shoemaker, the "beloved" principal of the school, patted her on the head and said kindly: "Thee expresses thyself well, my child. Be careful that thee does not let thy imagination run away with thee."

Agnes spent two years at Friends Central, 1888–90, during which time she mixed picnics and tennis with her studies—and, according to her mother, became "a full fledged Unitarian." At the end of each school term, she eagerly returned to New Mexico, where she exchanged her city clothes for the five-gallon Stetson and rugged skirts she wore in the Datils.

As she progressed into her teens, Agnes assumed new responsibilities on the ranch—and, in fact, did the work of a regular ranch hand. She learned to shoe horses, rope calves, and drive horse-drawn wagons into Magdalena for supplies. She assisted at roundups, two being held annually, one in the spring to brand the new calf crop and the other in the fall to segregate cattle to be shipped east. Although she "rode sidesaddle like a lady," she later recalled, "the double standard did not exist on the ranch." She worked "side by side with the men, receiving the same praise or same censure for like undertakings." And she did her share of "brush-breaking," riding swiftly into the pines after runaway cattle, breaking off branches with the momentum of her body.

Some tasks she assumed were exhausting and not without danger. She long remembered the time she set out before daybreak to deliver a saddle horse to a cowhand who had lost one of his string. Under a full moon, she suddenly discovered a pack of coyotes following her. When the cowboy failed to appear at the rendezvous site, she was forced to ride all the way into Magdalena, leading the extra horse—the coyotes following her most of the way. Although she knew that coyotes never attacked a rider, their howling nearly unnerved her nonetheless.

On another occasion, when Agnes was staying at one of the outlying camps looking after horses in a nearby pasture, a small herd of cattle with a half-dozen cowpunchers went by her cabin. The men stopped for a drink of water and discovered that she was alone. After the outfit departed, the boss-owner returned, entered the cabin, and, seating himself beside the teenager, "reached over and ran his cupped hand down the length of my braided hair." This was the only time, she wrote in her memoirs, that she felt "even vaguely ill at ease when I found myself alone with any man." Yet, with remarkable composure, she laughed, crossed to a side table upon which lay her small thirty-two, stuck it into her belt, and returned to his side to continue their conversation. He left, muttering, "I never let no woman take a shot at me." Later, Agnes avowed: "We had a saying, 'A six-shooter makes all men equal.' I amended it to 'A six-shooter makes men and women equal.'" By coping with these and other difficult situations, young Agnes acquired an "extraordinary self-assurance," a trait that helps account for her independent lifestyle in later years.

The summer that she turned sixteen, "Miss Agnes" had a different sort of adventure when she became the local schoolmarm and attempted to impart the three Rs to twelve youngsters, two under the age of six and one about her own age. A temporary replacement for the previous teacher, Agnes agreed to take on the job, although it meant riding fourteen miles round trip on horseback to the schoolhouse. The log cabin where she was to teach lacked many of the amenities found in city schools, such as desks, books, and piped-in water. One resident recalled that "the children drank from a bucket with a common dipper and sat on chairs or boxes." At the end of the term, when the county refused to pay Agnes because no contract had been signed, the parents took up a collection—and she received twenty-three dozen eggs for her efforts.

While in her early teens, Agnes became aware that "all was not going as well financially as might be" on the ranch. On her first summer home from Philadelphia, she and Ray discussed the matter, concluding that their mother was "tragically miscast as a range boss." Thereafter, as Agnes wrote in her memoirs, the two oldest Morley children "[began] to take over the management of the ranch. . . . [and Mother] gradually let us do it when we weren't away at school."

Still, the young Morleys had the help of some capable cowhands and also of Ada's cousin Orrin McPherson, an Indiana "farm-bred" young man in his twenties. Orrin fails to make an appearance in *No Life for a Lady*; but in an unpublished essay Agnes wrote that Ada had invited him to help run the ranch about the time that Floyd Jarrett left the family. The exact date of Floyd's departure is not known for certain, but he probably started east on a business trip in 1889. Although Agnes's mother corresponded with Floyd for at least a couple of years after that, most likely he rarely (if ever) again set foot in the White House. Thus, Agnes poignantly described the situation after her stepfather left: "[Mother] found herself marooned with three young children on a desert island of cultural barrenness with no means of escape that would not sacrifice her entire investment. We became a sort of Swiss Family Robinson without a Father Robinson to meet emergencies."

Life must have been difficult for Ada's family during the last decade and a half of the nineteenth century, for hard times had fallen on much of the nation's western cattle ranges. The good years in the stock business had ended soon after they arrived in the Datils; cattle prices had already started to slip and would continue to slide downward for several years. By the spring of 1889, one historian contends, the country was "in the midst of a severe cattle depression." The *Las Vegas Stock Grower* reported on the gravity of the situation: "Many stockmen have been completely ruined, while others have been forced to give up the business of a lifetime with great financial loss" (January 4, 1890). Then, too, periodic droughts and at least one severe blizzard during these years spelled disaster for many Socorro

County ranchers. Most likely, the decline in Ada Morley's ranching interests stemmed more from these economic and climatic conditions than from an inability to cope with cattle rustlers, as Agnes suggested in her memoirs.

In the midst of her financial troubles, Ada embarked upon a crusade to recover some of the Morley estate that rightfully belonged to her children. Agnes touched upon her quest in *No Life for a Lady*, recording that when her father died, he left a will executed before the birth of any of his children, bequeathing everything to Ada. Some years later Ada discovered that the will was invalid precisely because it failed to provide for his children, who legally were entitled to part of the estate. By the time she made this discovery, she had given Jarrett a power of attorney to sell a parcel of land that Ray Morley had owned in what is now downtown Denver, Colorado. Testimony would reveal that Jarrett sold the property for a fraction of what it was worth. On the advice of Smith McPherson, Ada's cousin (and Orrin's brother), who practiced law in Iowa, she secured a court order in 1889 appointing Asahel M. Andrews, a Denver attorney, as the guardian of the children; Andrews then spearheaded the drive to recover their interests in the Denver real estate. A long, drawn-out legal battle ensued, involving a succession of lawyers, guardians, and judges, to make certain that Agnes, Ray, and Lora had sufficient money for their support and education. Agnes best summed up the situation in these words: "Twenty years . . . were required to settle all the claims involved, and the 'Denver lawsuit,' as we called it, became a vexation of spirit, though in the end the sole source of family income."

Ada would spend a good deal of time and money traveling to Denver to protect her children's interests. On occasion, one or more of the youngsters accompanied her; but when she set forth alone in August 1890, she engaged a local woman to cook and look after them. Ada's absences undoubtedly contributed to their independent spirits and self-sufficiency. While her mother was away on this trip, Agnes felt duty-bound to ride several miles to help her friend, Lulu Starring, "hold down the [family] ranch," as she put it. When Ada's stay in Denver extended beyond her original plan, she lost track of her two girls. "No news is good news," she wrote to Manley Chase. "Not a scratch of a pen from them for 10 days. They are at our home ranch. Have horses, ponies, and freedom from all conventionalities so I trust they are happy."

Ada kept the children on the ranch the winter of 1890–91, explaining to the Chases that Agnes needed a "home life and home influences and a mother's guardianship. She is of course intellectual but a year of home would do her good. . . .We are all very happy a reunited family here after years of separation." But Ada constantly worried about expenses. In November she asked Manley if he knew of any cattle buyers. "I must sell some cattle to pay this summers expenses. The cowboys haven't been paid and are kicking up a muss." Still, by the end of the year (1890), she had contracted to hire a male tutor for six months, who, as it turned out, was a whiz in math but highly disliked by the children.

The future looked bleak for Agnes's family, as debts piled up and the mortgage to G. N. Gentry hung over their heads like the proverbial sword of Damocles. Sick from worry, Ada wrote to Floyd Jarrett in February 1891, telling him "to come out here and gather up the tangled ends of business." It seems, however, that neither Ada nor Floyd had the resources to meet Gentry's demands, and the Texas cattleman foreclosed on the ranch on July 1. When Agnes learned that "the house, range, cattle, horses, etc." would go to satisfy a $9,000 debt on the cattle, she solicited Manley Chase's help, urging him to suggest a plan whereby she, on behalf of her brother and sister, could raise the money to save the property. Aware of the legal problems involved, she wanted the ranch placed in the children's names "to keep it out of Mr. Jarrett's control."

County authorities soon descended upon the Jarrett ranges to gather the cattle and horses to be sold at a public auction. In *No Life for a Lady*, Agnes vividly described how she rescued her pet pony, Gray Dick, from the clutches of a deputy sheriff who had badly mistreated the animal during the roundup. What happened next to the Morley-Jarrett property is uncertain, for county records fail to reveal the names of any buyers. Agnes stated, however, that "the ruin I had expected never quite came. Money dribbled in from Denver, the ranch did not have to be sold, some of the cattle and horses were salvaged, and life went on, not smoothly, but certainly not with the drabness I had feared."

In fact, Agnes's life soon entered an extremely difficult phase, although in *No Life for a Lady* she never alluded to the events that were about to unfold. In August, she returned to Philadelphia to enroll in a private school operated by a Mrs. Lucretia M. Mitchell. Short of funds, she again turned to Manley Chase for help. "I'm in a horrible situation," she cried. "My school tuition is due, besides other expenses . . . [and Mama] cant send it to me." Nonetheless, the family scraped together sufficient funds to allow Ada, Ray, and Lora to join Agnes in Philadelphia for the winter.

In January 1892, Agnes's mother received a long, thought-provoking letter from Chase, now the legal guardian of the two younger Morleys, urging her to return to New Mexico—no doubt to discuss recent twists in the Denver lawsuit. But Ada's floating kidney was bothering her a good deal, so Agnes volunteered to go in her place. A mature young woman at the age of seventeen, she had the complete trust of her mother, who did not hesitate to empower her daughter to act in her stead on business matters. Ada, of course, could not foresee the turn of events that threatened to change the direction of her daughter's life in an alarming way (at least in Ada's eyes).

Agnes Morley arrived at the Chase ranch sometime after March the first. On April 23, she and Mason G. Chase, the twenty-two-year-old son of Manley and Theresa, were married at the home of Mason's parents. The details surrounding this unexpected development remain a mystery. Agnes never wrote about it; and two conflicting accounts of what transpired after the wedding—one put forward by Norman Cleaveland, Agnes's son, and the other by Mason Chase—add to the confusion.

Norman Cleaveland has questioned whether the marriage was legal in the first place. Agnes was too young to be married without parental consent, he points out, and the preacher who signed the marriage certificate identified himself only as an elder of the Methodist Church. Still, the trickiest part to unraveling the mystery is understanding Agnes's emotional state or motives at the time of the ceremony. Cleaveland rejects the story concocted by Ruth Armstrong in *The Chases of Cimarron* (1981), a story that has the handsome Mason Chase sweeping Agnes off her feet. In contrast, Cleaveland believes (based on cryptic remarks found in Ada Jarrett's letters to the Chases) that "the alleged marriage was merely a ploy in a highly involved plot to plunder the Morley estate" and that Agnes played along for her own reasons.

Moreover, Chase descendants have told Cleaveland that upon the completion of the ceremony, Agnes tossed the wedding ring into the garden and retired singly to her room. Ada, who had been warned by Cimarron friends of the pending marriage, stormed into the residence the next morning, furious with the Chases for allowing this travesty to take place—thereby jeopardizing her daughter's chance for an education—and removed Agnes from the premises. She soon started legal proceedings to have Chase dismissed as guardian for Ray and Lora.

Ada's bitterness toward the Chases, once considered among her closest friends, overflows in her letters to Manley. "You and your wife designed and executed that midnight mockery," she fumed, calling it an attempt "to steal Agnes" in order to control the Morley estate. But her daughter won out in the end: "You found you could not keep 'the bird you had caged,' as Agnes expresses it to her lawyers and friends." Ada continued to trust Agnes implicitly. "I have Agnes Morley's word for all I know of the villainous scheme and that satisfies me. She never in her whole life told a falsehood—nor practiced a shadow of deceit. I have absolute faith in her and her truthfulness." Norman Cleaveland believes that a lawyer "assured [Ada] that there had been no legal marriage"; therefore, there was no need to seek a divorce or an annulment.

A different story emerges from a deposition Mason Chase gave two years after "the midnight mockery" when he sought, and obtained, a divorce from Agnes on the grounds of abandonment. He claimed that he and Agnes had lived together "as man and wife" for about three weeks before her mother arrived and took her away. He also claimed to have made frequent appeals to Agnes to return, but she refused. Interestingly, the Notice of Publication for the divorce appeared for six consecutive weeks, as required by law, but in a Spanish language newspaper published in Las Vegas. A legal notification also was sent to Agnes at her last known address (she was then living in Ann Arbor, Michigan), but she failed to respond, if, indeed, the document ever reached her.

And so the mystery remains unsolved. But it is clear from Ada's letters to the Chases and Mason's deposition that, whether he and Agnes cohabited as man and wife after the wedding ceremony or she stayed chastely in her room, Agnes Morley wanted nothing more to do with Mason Chase.

In all likelihood, Agnes spent the fall of 1892 in Ann Arbor, where she finished high school. She probably had not escaped the Chase affair unscathed emotionally; Ada hints at her daughter's state of mind in a letter to Manley Chase, written on March 25, 1893, a letter filled with accusations as well as the news that Agnes was in school: "Of her own volition she went to school but it took all her strength and my help to defeat your sinister plans."

Agnes made only brief mention of this school year in *No Life for a Lady*: "At Ann Arbor as at Philadelphia life had very little reality for me. I worked hard, but everything except the ranch was unimportant to me. The next summer found me joyfully getting off at the Magdalena railroad station." In fact, Agnes's buoyant personality must have saved her from falling into a morass of self-doubt and depression. Within a year of the "midnight mockery," she seems to have erased it from her memory. The very summer that she left Ann Arbor, she traveled to Chicago with her mother and sister to take in the marvels of the World's Fair. They stayed with Ada's relatives, the Charles E. Tibbles family, part of the Tibbles clan that included Ada's mother, Mary Tibbles McPherson (an older sister of Thomas

Henry Tibbles, the crusading journalist who worked for Indian rights). Walter Owen Tibbles, Charles's son, later wrote to congratulate Agnes on the publication of *No Life for a Lady*, stating that he still retained fond memories of her visit nearly fifty years earlier. "I have not forgotten the thrilling descriptions you gave us of ranch life at that time," he added.

Agnes returned to Ann Arbor in the fall of 1893 to enroll in the University of Michigan, one of the nation's largest universities, with a student population of about three thousand. By the end of the decade, 40 percent of Michigan's undergraduates would be women. Indeed, the nation had witnessed a dramatic increase in the number of females enrolled in institutions of higher learning since Agnes's mother had attended an Iowa university in 1870, jumping from eleven thousand to eighty-five thousand in 1900. Still, Agnes and her cohorts were rarities, for, as Barbara Miller Solomon points out in her book *In the Company of Educated Women*, only 2.8 percent of the nation's young women went to college at the turn of the century.

Agnes entered Michigan by examination, meeting all requirements except for physics, a "condition" which she soon removed. In the following two years (1893–95), she took a large array of classes—ranging from Latin, French, and German to history, philosophy, political economy, and mathematics. A single course in mineralogy may reflect an interest she acquired while riding over the rocky hills of Socorro County.

In her younger days, Agnes aspired to become a teacher, but evidently she did not pursue this goal while attending the university. Nor was she the "lady scholar," as described by Barbara Solomon. Rather, she fit the mold of the "all-American girl," entering wholeheartedly into campus life—she joined the Alpha Phi sorority, served tea to faculty and their wives at social gatherings, and took part in athletics. One female classmate remembered her as being "witty and attractive."

Another friend, Arthur Kendall, who lived near the university with his family, described Agnes as "the most accomplished horse-woman in the whole wide world." And because of her healthy, outdoor, all-American girl personality, he may even have developed a crush on her. Years later he wrote: "The Peckham girls could never understand why I took to you more than I did to Katy. . . . Katy was a grand girl but it was only when with you that I could breathe freely from the great outdoors, and I liked the outdoors and the great open spaces."

Arthur remembered the first time he allowed Agnes to ride Maud, one of his two horses. Afterwards, the president of the university, James B. Angell, begged him "never to let that girl ride either of those vicious horses again"—but withdrew his objections after he learned that Agnes came from a cattle ranch in New Mexico and could shoe her own horses.

Another of Arthur's remembrances is well worth recording, for his story corroborates those Agnes told in *No Life for a Lady* about her equestrian feats:

> Once, when you were riding Maud, and she could not keep up with Black Beauty, I looked back to see the fun and the fireworks, for I knew Maud well. Her ears were back flat and there was murder in her eyes. She was mad all the way through because she could not lead in the race, and when Maud was mad it was well to watch your step. As she was passing a farm house she turned right, almost at a right angle, and made for a closed gate with the evident purpose of jumping the gate. Although I never knew a horse to make a quicker turn, I don't believe your legs (limbs in those days) left the side of the horse enough to shove in a piece of paper between them and the side of the horse. And you had the little devil under complete control before she reached the gate. You acted as if that was an everyday occurrence.

Agnes spent the summer of 1894 on the ranch, along with her brother Ray, who had been attending the Chester Military Academy in Chester, Pennsylvania. Both Morleys would long remember the danger, excitement, and high-jinks of this particular summer. At the age of eighteen, Ray was a young man with a vision and a mission. To advance his dreams "of getting the ranch back on its feet," as well as expanding its operations, he had entered into a partnership with an older local resident. While at school, however, he had learned that his partner planned to sell their cattle and "skip the country with the proceeds." Ray was determined to reclaim his half of the herd before that happened. In her memoirs, Agnes told how, on the second night of her homecoming, she and Ray wrapped their horses' feet in gunny sacks, rode silently to the pasture where his cattle were corralled, and then quietly pushed them through a lower gate and spirited them away—with four armed men playing cards in a nearby cabin. Had they been caught, Agnes surmised, they would have been shot, "for we *were* trespassers—we *were* cattle rustlers!"

Not long after that escapade, Agnes and her mother were alone on the ranch when two young strangers appeared and asked for lodging. Their demeanor frightened Ada, for she connected their presence with the recent brutal murder of a neighbor. Agnes, suspecting that they intended only to steal some of the Morley horses, devised a novel means of foiling their plans. She exchanged wisecracks and played poker with them through the night, and away they went the next morning in good humor. It was later confirmed that the strangers were indeed outlaws on the run.

Still later that summer, the ranch was the scene of much merriment, when Agnes's sorority sister Grace Ward arrived for a lengthy visit. Apparently, she was the first in a long line of college youths who made their way to the Morley home. Agnes later was to write (in an unpublished essay) that the ranch "saw its most colorful era" during the college days of the young Morleys. "Every summer vacation saw it converted into a summer resort for our schoolmates. . . . One summer established a record of twelve different colleges being represented in the persons of our guests."

Grace would never forget her stay at the ranch the summer of 1894. Some of the Alpha Phi sisters had taken her aside before she left for Datil and said to her: "Grace, we want you to take careful note while you are visiting Agnes. You know, Agnes worries us. She seems to be a truthful enough person generally, but we just can't believe all the things she tells us about that

New Mexico cattle ranch. Just keep your eyes open and report." Not long into her visit, Grace responded: "I want to report that Agnes' stories indicate admirable self-restraint on her part."

After publication of *No Life for a Lady*, Grace Calhoun (her married name) wrote to Agnes, recalling the time that one of the cowpunchers asked if women really did wear clothes like the pictures (of evening gowns) in the magazine that Grace was reading. Thereupon Agnes, Grace, and Grace's friend Elizabeth decided to dress up for supper. But, Grace wrote, "Elizabeth gave it away (as usual), and the boys came in with spurs, guns, hats and full regalia." Ray had planned it all, she recalled. And when he drew his gun "to shoot the coffee pot off the table to help us clear up . . . [I] remembered your instructions. I went under the table. Bob's spur caught in one of my ruffles and he walked off with yards ripping after him. . . . Elizabeth fainted and all was forgotten in bringing her around." University life must have seemed pretty tame after a visit to the Morley ranch.

Another Michigan classmate, Pearl Colby, also contacted Agnes after publication of her memoirs, and reminded her of events surrounding the opening of the new Waterman Gymnasium in October 1894. Women were allowed to use the facility only in the mornings and men in the afternoons and evenings. "There were two outstanding girls in the gym work," Pearl recalled, Helen Coe and Agnes Morley. Pearl simply tagged along and "tried to be outstanding." The threesome would go in the afternoons to watch "the boys' stunts" and then "nearly break our necks the next morning, trying to do the same things." Pearl also remembered how they helped Mrs. Campbell, wife of the blind chemistry professor, E. D. Campbell, learn to ride a bicycle, so "she could ride a tandem with her husband. We three took turns, two at a time, helping her, and she was no light weight. We trotted round and round that gym by the hour, holding her up."

Pearl then recounted an incident that gives a glimpse of the trio's budding feminism. When they were told that regent Levi Barbour was giving $20,000 to build a gym for the girls and that a formal announcement would be made in the auditorium, they decided to give an "appropriate yell for Mr. Barbour" to show their appreciation. Being the only senior, Pearl was chosen to lead the yell. But word got out, and the editor of the university's daily newspaper protested against this "unladylike" behavior. Finally, after consulting with Fitzgerald, their gym instructor, who advised against the yell "on the ground that gym work for women was on trial as yet," the young women "said not a word," but "waved our kerchiefs when the announcement was made."

Agnes chose not to return to Ann Arbor to begin a third year at the university but instead remained in the Datils to help her mother run the ranch while Ray went to college. Thus, at age twenty-one in the fall of 1895, she had one of the most memorable experiences of her young life—hunting grizzlies with Montague Stevens, a transplanted Englishman and big-game hunter who lived west of the Morleys. Two years previously, Stevens had as guests on a bear hunt General Nelson A. Miles and the artist Frederic Remington. Stevens would later win acclaim as author of *Meet Mr. Grizzly: A Saga of the Passing of the Grizzly* (1943).

In *No Life for a Lady*, Agnes described in detail the ten-day hunt, conducted from the Morley ranch as a base of operations. She carried no firearms and went along merely for the adventure, while Stevens and his friend Dan Gatlin shot and killed two grizzlies, one black bear, and "jumped" four others. With her superb writing ability, Agnes recreated for her readers the sense of danger and excitement that accompanied the three hunters when they set out after a grizzly at a dead run across rugged terrain and then, finding themselves marooned on a narrow shelf, lowered their horses by rope down a steep cliff. When one horse was injured Agnes urged the two men to continue the chase while she staggered home on foot with the horse in tow, arriving about one o'clock in the morning, "benumbed with cold, my clothing in tatters." Danger also lurked the day Montague fired into a bear den while Agnes perched in a nearby tree. Dan chivalrously gave Agnes the opportunity to fire his rifle when the bear emerged. She did not kill the bear but feared she had shot Montague when he tripped and fell at the exact moment she fired.

With little ranch work to be done in the winter, Agnes decided in January 1896 to enroll in Leland Stanford Junior University, a new institution in its fifth year of operation. The university's rural setting, on the Stanford estate in northern California, must have pleased the young woman from the Datils. So, too, would she enjoy the congenial relations between the faculty and student body, which numbered about 1,100 (approximately 40 percent women). Only part of the students could be housed in Stanford's two dormitories (Encina for the men, Roble for the women); the overflow, which may have included Agnes, found lodging in the farm towns of Mayfield and Menlo Park, in newly built Palo Alto, or with faculty families living on campus.

Agnes declared economics as her major, seeking the knowledge she felt she needed to help run the business side of the ranch. During the five semesters she spent at the university, she took a majority of her classes in the department of economics and sociology and a smattering of other courses as well, including English (the only required course at Stanford), nineteenth-century poetry, French, German, Spanish, zoology, and hygiene (gym). In her senior year, much to the horror of her sorority sisters, she signed up for a blacksmith course, offered by the mechanical engineering department, to brush up on her technique of shoeing horses.

Like other college students, Agnes absorbed many of the social and political ideas of her faculty mentors, especially those of Dr. Edward A. Ross, who taught courses both in economics and sociology. A popular teacher, he was known for his progressive ideas and for stimulating his students' thinking. But Ross's affinity for free-speaking, both in and outside the classroom, raised the hackles of Jane Stanford, who, along with her recently deceased husband, Leland, had built the university. She came to believe that Ross was a socialist, with dangerous ideas, "bent on covertly spreading his views in his teachings." At her insistence, university president David Starr Jordan, in

1900, asked for, and received, Ross's resignation, an action that created a national furor and lengthy debates over faculty tenure and freedom of expression.

Ross went on to have a distinguished career as a sociologist. But, as Mrs. Stanford had feared, he had had enormous influence on the thinking of Stanford students. Agnes Morley, in fact, became one of his disciples, and, according to her son, would register as a socialist when she had the opportunity to vote in 1911. No doubt it was for Professor Ross's class that she turned in an assigned essay on the meaning of civilization. The thesis she developed was one she reiterated for several years, that "civilization is the gauge of the extent of protection given the weak by the strong." She would later have a change of heart and adopt a more conservative line.

With her gregarious nature, Agnes seems to have led an active social life on campus and to have acquired a host of friends. She took part in sporting events and theatricals, went on picnics to Pacific Grove near Monterey (a favorite destination for students and faculty), and helped establish a local chapter of Alpha Phi sorority. She counted among her friends Lou Henry, who loved the outdoors as much as Agnes. Lou had entered Stanford in 1894, met Herbert Hoover in a geology lab, and by the time he graduated the following spring, the two "shared an emotional commitment." She graduated in 1898 and married "Bert" the following year. There followed years of travel, public service, and Herbert Hoover's election to the presidency. After the Hoovers left the White House, Agnes and Lou Henry would work together to advance Republican causes. A letter that Agnes wrote to the former first lady in 1934 provides a clue as to the depth of their college friendship: "It was a very great joy to have enjoyed your hospitality [at the Hoover residence] and to have renewed a friendship that has never been dimmed in our minds in face of the most severe of all tests, the rise to eminence of one of the parties while the other remained in the ranks of modest citizens."

Agnes also formed life-long friendships with Mary Roberts Smith, a young professor in Stanford's sociology department, and with Dane Coolidge, one of Smith's students, who went on to become a writer of western fiction. Mary and Dane were married in 1906, after she had divorced her first husband and left Stanford. In later years the Coolidges often visited the Morley ranch, where Dane soaked up atmosphere for his novels. Writing to Agnes after publication of *No Life for a Lady*, Mary Coolidge admitted: "I am not able to talk about it dispassionately . . . because it brings back to me the little group whose lives became entangled more than forty years ago at Stanford, who have remained friends ever since though often far apart."

Other school chums remembered the stories Agnes told about the Morley ranch. After Sarah Morrison finished reading Agnes's memoirs, she wrote to her friend, "My dear Sapiens" (a form of address they had used in college), "[I] feel that I've just been back to a college reunion and had a long talk with you. As I read I could just hear you say *Corky* [a cowboy] and *Montague Stevens* as plainly as when I heard you tell about them over forty years ago."

Agnes gained her greatest notoriety on campus during her first semester, however, when she played on the women's basketball team and scored the winning basket in a game with the University of California. Nothing in the records indicates when Agnes was introduced to basketball, a new sport invented in 1891. The first collegiate women's basketball game was played at Smith College in 1893, and the game quickly spread to other colleges. In the early days at Stanford, sports for women had been encouraged, with the emphasis placed on exercise and good health, not competition. But with the backing of the Women's Athletic Association (of which Lou Henry was a member), competitive sports would become standard.

The contest with the Berkeley women, the first women's intercollegiate basketball game in the United States, was played in the armory in San Francisco on April 4, 1896. Seven hundred spectators, all women, jammed the gallery. Men had been barred in deference to the Berkeley women, who had never appeared in mixed company in their gym suits (unlike the Stanford women who regularly practiced in view of their male classmates).

Mary Roberts Smith chaperoned the team and composed a special yell for the game, which, the press reported, "the Stanford women rooters gave with great effect." When news of the women's victory reached the Stanford campus, the men of Encina Hall staged a noisy impromptu celebration in the hallways and then joined other admirers to cheer the players when their train reached Palo Alto at 8:00 P.M. An omnibus decorated in Stanford colors took the team to Roble Hall, where the women residents "carried right guard Agnes Morley into a reception room and made her tell the story." The next day all three San Francisco newspapers ran detailed accounts of the contest.

Agnes Morley continued to be active in collegiate sports in following months, as did Lou Henry. In the fall of 1896, the Women's Athletic Association held its annual election, with the results announced in the *Oakland Tribune*: Mrs. David Starr Jordan, president; Lou Henry, vice-president; and Agnes Morley, treasurer.

Agnes left the university for the 1898–99 school year, probably staying on the ranch to help her mother. When she returned in the fall of 1899, she moved into the Alpha Phi sorority house in Palo Alto. Her days were filled with activity—class work, sorority dinners and dancing parties, and . . . preparing for her wedding.

In *No Life for a Lady*, Agnes said very little about her husband, Newton Cleaveland, other than that they had met on the Stanford campus and that he never adjusted to life in the Datils. Newton was about the same age as Agnes, having been born in the Province of Quebec, Canada, on February 6, 1874. Later the family moved to California, where Newton attended public schools in Butte County. When he entered Stanford the fall of 1894, he majored in physiology, intending to become a physician like his father. Once he left Stanford, however, he would make a name for himself in the nascent California gold-dredging industry.

Almost nothing is known about the courtship of Agnes Morley and Newton Cleaveland. A newspaper clipping from 1944, however, reveals that they had been part of a Stanford group that made Pacific Grove "their headquarters for oft-repeated visits, and picnicking on the beach." It is easy to imagine that they attended many other social events together as well. At any rate, Newton graduated ahead of Agnes, in the spring of 1899. She completed her course work just before Christmas, and they were married in Trinity Church in San José on December 23. But Agnes would have to wait until the spring commencement of 1900 to receive her degree (she liked to say she graduated in zero-zero or "aughty-aught").

Like many other couples, the Cleavelands would encounter rocky patches in the days ahead, yet their marriage endured for nearly forty-five years, possibly because they made allowances for each other's idiosyncrasies. Newton Cleaveland must have known he was marrying a headstrong, independent-minded woman—one whose identity was undeniably tied to the New Mexico landscape. Two days after the wedding, Agnes inscribed a five-stanza poem to her new husband on the back of a photograph showing her in a frilled white blouse. The poem proclaimed her prowess as a cowhand, as well as her love of the Datils. A sample follows:

> Would she ever chase an "oso" [bear],
> could she turn a hoolian [small-looped lasso],
> jack a shoe on her own pony,
> in the absence of a man?
> There you may behold her
> gowned in ruffles, puffs and frills
> but the heart that's underneath them
> is still in the Datil hills.

For Agnes, the Datil Mountains and the Morley ranch were enchanted regions that would tug on her heartstrings for the rest of her life. Time and again she found reason to return, even though she lived most of her married life in California. And it was there on the West Coast that she gave birth to four children and carved out a career for herself as writer, Christian Science practitioner, and activist clubwoman.

CHAPTER 3

A Jekyll-Hyde Life

FOLLOWING their wedding, Agnes and Newton Cleaveland took up residence in Oroville, California, where Newton had found work in the gold-dredging fields after graduating from Stanford. Although the Gold Rush era had long passed, Oroville and its environs continued to produce large amounts of gold in following years. Oroville's most glorious period as a gold producer, however, stemmed from the introduction in 1898, by Wendell P. Hammon (Newton Cleaveland's longtime employer), of the first successful dredging operation in California, working an area along the Feather River. In short order, the Oroville dredge field became one of the best-known and most profitable mining districts in the world.

Agnes left no record of how she spent her days when she lived in Oroville, then a town of perhaps two thousand residents, with imposing business buildings lining its principal streets. Being adventurous, she no doubt explored the surrounding countryside and became acquainted with her neighbors. Still, the California gold country could not compete with the Datil Mountains for Agnes Morley's affections. Newton recognized this fact quite early in their marriage, as suggested in the notation he wrote in a twelve-year-old girl's autograph album. He and Agnes had encountered the youngster in Socorro on January 2, 1901, and upon her request Agnes wrote her name in the book and her address as "Datil, New Mexico." On the opposite page, Newton wrote: "My wife's address is Datil, N. M. My address is Oroville, California."

After her marriage to Newton, Agnes frequently returned to the Datils, rarely allowing a span of two years to go by without a visit. On the Morley ranges, she resumed the role of a rancher, donning her five-gallon Stetson and helping with the chores. In *No Life for a Lady*, Agnes described this movement back and forth between the city and the ranch as her "Jekyll-Hyde" life. At the beginning of her marriage, however, her ranching inclinations appeared to be winning out over any allures that the city might hold for her.

Ten months after Agnes gave birth to her first child, Norman, on April 4, 1901, in Oakland, California, Ada summoned her daughter by telegram, saying there was an emergency on the ranch and she was needed. The real reason for the telegram, Agnes suspected, was that her mother wanted to see her first grandson. So the new mother traveled by train with the baby as far as Magdalena. There a young lad waited with a buckboard and team to drive them to the ranch. They set off early in the morning, Agnes well supplied with diapers, bottles of milk, and their baggage. This was to be one of Agnes's most memorable homecomings.

Half-way to the ranch, a coyote loped across the road in front of them. The inexperienced driver dropped the reins, grabbed a rifle, and, with Agnes shouting "Don't shoot, don't shoot" while holding the baby in her arms, he shot. The horses bolted, broke the traces, and away they galloped across the plains, the driver running after them. Agnes and baby Norman sat for three hours in the buckboard before a freighter came along and gave them a lift into Datil. But this may have been the worst part of the trip. Because the wagon was fully loaded, the passengers had to sit on top of a coop of chickens in the back. According to Norman, "the rocking of the wagon was sickening, the smell of the chickens was also sickening. But worst of all, the milk turned sour and I could not abide sour milk. So I nearly squalled my lungs out. Then Mom ran out of diapers, so she was in a horrible mess."

Once they reached Baldwin's, at 2:00 A.M., the proprietors fed and took care of them. The runaway team returned to the ranch before dawn, the tired driver arriving about an hour later. At daybreak, a worried Ada hitched up another team to go in search of her daughter and grandson, and was delighted to find them at Baldwin's.

Agnes's mother no longer lived in the White House, however. A few years earlier, when all three Morley children had been away at school, Ada took an extended trip, leaving caretakers in charge of the house. She returned to find it deserted and stripped of its small furnishings. That spelled the end of the White House as the Morley home. After a family council, Ada purchased property from Johnny Woods—known thereafter as the Swinging W Ranch—located up Main Canyon about three miles from the Datil post office. The house had been built by a Norwegian, "who understood cold weather construction." "Its double plank walls were packed with adobe and its rooms were small and low ceilinged," Agnes wrote in an unpublished essay, "but beyond all it stood on a sun-reached southeastern slope of hill instead of in the bottom of a cañon which got sun only in the midday hours."

Norman Cleaveland later recalled that he spent much of his childhood in the Datils, residing there off and on between 1902 and 1907. During this time, Agnes returned to California to give birth to Loraine on November 18, 1902, and to Agnes Morley (known as Morley) on June 26, 1905—Agnes's thirty-first birthday. Meanwhile, Newton Cleaveland pursued graduate work in mining engineering at Stanford in 1902–1903 and also taught physiology there as an instructor. After he joined Agnes in the Datils during the summer of 1903, the local press referred to him as "Prof. N. Cleaveland of the Leland Stanford University."

Newton probably spent much of 1903–1904 living with Agnes and the children on the Swinging W Ranch. On July 15, 1903, the Cleavelands bought 320 acres situated about three miles from the now-abandoned White House. This acreage included Jack Howard Flat, a place Agnes had coveted since a child. Although the Cleavelands made no improvements at this time, the property would remain in the family beyond Agnes's death. Newton tried adjusting to life on a cattle ranch but did not succeed. Agnes described some of his difficulty in these words: "When he had planned to use a team on the following day he expected the team to be available, which was more than the rest of us did. After he had wasted just so much time and energy looking for straying horses he decided that his time and energy could be better spent looking for gold mines." Agnes and Newton returned to California in February 1905—she to await the birth of her third child, he to join W. P. Hammon's dredging operations in Butte and Yuba counties.

Still, Agnes seems to have thrived during her days on the Swinging W. She had abandoned the sidesaddle in the mid-1890s and adopted a man's saddle and a divided skirt. A photograph taken on the ranch in 1903 shows her wearing a stylish outfit astride a horse, with little Norman seated in front of her. But it hadn't been easy to break with tradition. When Agnes approached a neighbor's home riding "clothes-pin fashion" instead of "sidewise," the outraged woman responded: "You aint welcome in my home no more. . . . I always thought you was a good girl." Her brother Ray also blurted out his disapproval: "I won't ride in the same cañon with you." But Ray and the neighbors soon dropped their objections, and Agnes took pride in having helped liberate ranchwomen from the "torture" of the sidesaddle.

Other photographs taken at the Swinging W in the early 1900s show Agnes helping to cut and stack hay. Occasionally, native grama grass grew in such profusion that ranchers like the Morleys rushed to harvest it. In her memoirs, Agnes described one season when her efforts ended in disaster. While she was "perched on the high iron seat of a rattle-trap mowing-machine, behind a supposedly gentle team of mules," they took off at full speed, frightened by the unaccustomed clatter of the machine. Although she regained control of the animals without mishap, she was less successful later in the afternoon when it came time to unharness the same team of mules. Working in tandem with Jesse Simpson, the foreman, Agnes failed to unhook the "second trace in time, and by some perverse instinct that mule team sensed that it was attached to the hayrack by a single strap of leather. . . . Away it went. Away went the hayrack with it, at first sidewise, then end over end"—the hayrack soon reduced to kindling. Jesse's only comment to Agnes at the end of the day: "Book-learnin' sure is a great thing."

Visitors continued to find a welcome at the Morley ranch in the first decade of the twentieth century. Agnes especially enjoyed the company of the young artist Frederick Winn, a recent graduate of Rutgers College. For a year or so, Winn worked as a cowpuncher on a nearby spread. His realistic depictions of western ranch life delighted a local reporter, who, with some journalistic license, avowed that in authenticity "he often surpasses Remington." The ranchmen of New Mexico so admired his work that in 1903 they presented some of his paintings to President Theodore Roosevelt during his visit to Albuquerque. Winn left the Datils to spend the winter of 1903–1904 at the Art Students' League in New York City. Upon his return in the spring, he took up residence at one of the Morley camps. A rare photograph of Agnes, probably taken in 1904, shows Winn standing at a rustic easel painting her portrait while she sits astride a saddle that rests atop a fence pole.

The arrival of a more troublesome guest, a woman writer of Wild West tales, would lead to Agnes's own career as a short story writer. In a humorous passage in *No Life for a Lady*, Agnes explained how, as a favor to a ranch hand, she ended up reading and critiquing one of the woman's narratives. Agnes quickly found that the author knew nothing about cow country and returned the manuscript "with exhaustive footnotes and reams of criticism." Pleased with these remarks, which she assumed came from the cowboy, the author telegraphed that she and her secretary would arrive in Datil within a week to confer further with him about the tale. Thereafter, the Morleys served as reluctant "source material" for the author. When her intrusive behavior became unbearable, Ray Morley "took matters in hand." To provide the writer with "local color," he staged a farcical buffalo hunt, using an old packhorse covered with a grizzly robe as the lone buffalo to be hunted by the ranch hands. The stunt was received rather coldly by the author, but, as Agnes reported, "she continued to sell 'Wild West stuff.'"

If this tenderfoot could sell a western story, Agnes reasoned, she ought to be able to do the same. Thus, a short story writer was born, and Agnes soon had several manuscripts in the mail to various magazines. There is no way of knowing how many stories she published in the days ahead; all of her records, including copies of manuscripts, were destroyed when the Cleaveland residence (in Berkeley, California) burned to the ground in 1923. Agnes's friend Mary Roberts Coolidge, however, claimed that she (Agnes) had written fifty or more pieces of fiction in those early years after she left Stanford. And a newspaper clipping from about 1935 states that Cleaveland had published "scores of stories" in *Munsey's Magazine*, *Cosmopolitan*, *Mother's Magazine*, *The Silhouette*, *Red Book*, and others.

Agnes seems to have broken into print in 1904, when two of her stories appeared in Philadelphia's *Sunday Magazine*, two in *Munsey's*, and one in *Metropolitan*. Justifiably proud when she sold her first story, she bought a buggy with her forty-dollar earnings. This may have been the piece entitled "Nine Points of the Law," which appeared in the August issue of *Munsey's*, one of the most popular illustrated monthly magazines in the nation. The story is based on an actual occurrence, when Agnes forcibly reclaimed her pony, Gray Dick, from the deputy sheriff sent to gather livestock for the foreclosure sale. Like so many other writers of westerns, Agnes mixed realistic portrayals of ranch life with a romantic plot, a sure-fire formula guaranteed to please readers at the turn of the century. The heroine, Belle Field, who "sat her horse like a young Indian," intends to solve her own problems without involving Clint Reed, a handsome but headstrong young cowboy who wants to marry her. To keep Clint occupied (out of respect for Belle's mission), friendly cowpunchers stage a night stampede, forcing Clint to help round up the cattle. Belle is therefore free to ride into the deputy's camp and take her pony at gunpoint. But rebelliousness is soon replaced by capitulation. When Clint arrives, Belle swoons: "Oh, Clint, I need somebody!" And she promised him "that never again would she try to fight one of life's battles without his help."

The second story Agnes published in *Munsey's*, "Chiquite" (November 1904), is also based on a real incident, one that involved a neighbor's child and that would appear without novelistic embellishments in *No Life for a Lady*. The hero of the tale is twenty-year-old Tierney, boss of a trail-herd that must be moved across a drought-ridden section of New Mexico. The cowboys accompanying Tierney are about to mutiny under the strain of riding herd on cattle that have gone too long without water. Finally, the trail boss rides ahead to a corral where the herd can be penned for the night—giving his exhausted men a chance for a good night's sleep. As the cattle rush toward the corral, its floor carpeted with "lamb's-quarter, that earliest of edible weeds," the cowboys spy a ten-year-old girl standing on the corral bars and a burro inside the enclosure. She waves her sunbonnet in the "faces of the bellowing animals nearest her" and cries out in a thin, childish voice, "You can't put your cattle in here! This is Yellowcat's pasture. You can't put your cattle in here, for it's all he's got to eat!" Touched by the child's agitation, the men agree to "night-herd" the cattle, leaving the lamb's-quarter to the burro. Eight years pass. Tierney returns to the area and discovers that "Chiquite" is grown up and managing her deceased father's ranch. By the end of the tale, matrimony is a foregone conclusion.

Agnes is at her best describing range conditions and the actions of the cowboys that ride herd on the cattle. She writes, for example, "It was one of those cruel drouth years of the Southwest, when hot sun and dry wind, day in and day out, week in and week out, bake from the earth every vestige of life-giving moisture, and the cattle grow slowly weaker and leaner, and at last give up the struggle." And she adeptly plays on the reader's emotions when she describes Chiquite's defiance of the hungry and thirsty steers. But the romantic plot tacked on at the end undercuts the story's powerful realism.

One of Agnes's strongest stories, "The Tramp Herd" (*Cosmopolitan*, March 1905), captured the essence of the cattleman's fight with the sheepman. It also featured a strong female protagonist, the newly married Katherine Halloran, who initially defied the imperious commands of her rancher-husband. Preoccupied with "a tramp sheep-herd" that was skirting the northern boundary of his range, Jim Halloran had left for the day, tersely admonishing her "to keep off all of those young horses." Feeling rebellious and proud of her reputation as "the best horsewoman in the county," Katherine lassoes and saddles Gruyer, one of the horses Jim had forbidden her to ride. Flying over the countryside astride Gruyer, she "exulted in her defection." Ten miles from the ranch house, she came upon a lamb in distress, which created a dilemma. She knew she could not dismount and then remount with a lamb in her arms, nor could she leave it behind. She chose to dismount; Gruyer shied away from the lamb, jerked the reins from Katherine's hand, and galloped away.

As the story unfolds, she meets up with the "Mexican" herder who has trespassed on Jim's range. She tells him to get off Halloran's land; "he's dangerous when he gets angry," she warns. The herder agrees to leave but only after she allows him to water his sheep at one of Jim's finest watering-places. Now Katherine faces the prospect of Jim's wrath. But he has been in agony since the riderless horse appeared at the ranch. Once reunited, the couple embrace, and Jim says huskily, "Oh, darling, I've suffered the torment of the damned for the last hour." "And Katherine knew that never again could such a thing as 'disobedience' be possible."

Among the strengths of this story are Cleaveland's realistic portrayal of a rancher's antipathy for sheepmen and her vivid word-pictures of the countryside. The romantic elements of the tale are weakly developed, however. And the surrender of a strong-minded woman to a domineering husband reflects the dictates of writing for a popular magazine rather than the realities of Agnes's own marriage.

Munsey's published another of Cleaveland's stories, "The Loyalty of Stephen Stovel's Widow," in April 1905. The storyline came from a tragic incident that took place in October 1903, when Corky Wallace, a cowboy on the Swinging W Ranch, shot and killed (in self-defense) a Socorro County rancher. The local press covered the murder investigation in detail, depicting both men as reputable cattlemen. In her short story, Agnes altered some of the known facts of the case. For example, she characterized only the protagonist as an upstanding citizen—the man he killed as a troublemaker.

It is not known what reception Agnes's story received in Socorro County, where residents had taken a keen interest in Corky Wallace's murder trial. When Agnes returned to New Mexico in August 1905, the *Albuquerque Citizen* merely noted her arrival and her recent publication: "Mrs. N. Cleaveland of Datil, N. M., who is one of the best known magazine writers of the day, is in the city, the guest of friends. 'The Fidelity of Steven Stovel's Widow' [*sic*] is one of Mrs. Cleaveland's stories which was widely read in New Mexico."

Agnes resided much of the time between August 1905 and February 1907 on the Swinging W Ranch. Most memorable during these months was the unexpected arrival of her brother Ray with his new wife, the former Nancy Brown, "the daughter of wealthy and prominent New Yorkers." Ray had led a Jekyll and Hyde existence also after he left the family ranch to attend school, first the Chester Military Academy, then the University of Michigan, and finally Columbia University, where he became a football hero. After he graduated with an engineering degree in 1902, he married Bessie Cresson, a short-termed marriage that ended in divorce. He coached the Columbia football team for two seasons (1902 and 1903) before joining a New York mining and engineering firm. During these same years he expanded the family land holdings and purchased more cattle. After he quit the engineering business (about 1908), ranching claimed his full attention and he eventually put together one of the largest ranches in west-central New Mexico—through land purchases, forest permits, the use of unappropriated public domain, and the financial aid of his wealthy mother-in-law.

Agnes was to meet Ray's bride for the first time on a bitterly cold February morning in 1906, when she set out to see a dentist in Magdalena, with Fred Winn driving the buggy. They reached Baldwin's before sunrise and spied a fancy-looking new buggy in a nearby corral. An employee on the premises told them it belonged to Ray Morley, who had arrived during the night and taken the end room in Baldwin's building. Fred bounded out of the buggy, pounded upon the door indicated, and yelled, "Hey! . . . C'm out of there you towheaded son of a hornswaggled sardine before I kick the door down." A faint feminine shriek was the response. "Fred Winn did not hear it," Agnes reported. "He was too preoccupied with banging on the door and threatening to shoot through it if Ray didn't open up immediately." But Agnes heard it. "It was my introduction to my new sister-in-law."

Two days after depositing his bride on the Swinging W, Ray took off for Chihuahua, Mexico, where he had contracted to inspect a mine. The three months of his absence, Agnes avowed, "were months of discipline" for both Ray's family and Ray's wife, who found that "money didn't buy immunity from inconvenience." By the time of Ray's return, however, Nan had made modest adjustments, even following the Morleys' example of ordering goods from the Montgomery Ward catalogue rather than from Sak's in New York City. Ray and Nan Morley soon took up residence at the Drag A headquarters, one of Ray's recently acquired properties located a few miles north of the Swinging W. In due time, Nan gave birth to triplets, Faith, Hope, and Charity, but only Faith survived beyond infancy. In following years, Nan made frequent visits to California and to her family's home in New York, and Faith attended boarding schools. For mother and daughter, the Morley ranch "served principally as a place in which to spend vacations."

Agnes's sister Lora also lived on the Swinging W periodically during the first decade of the century. In her Jekyll-Hyde life, she had briefly attended Stanford University but dropped out for medical reasons. She then spent a year in Germany visiting her Aunt Mary Schaper and teaching conversational English to adults in the Berlitz School of Languages in Berlin. After she returned to the Morley ranch, she met Perry Warren, a graduate of Yale who had traveled west for his health. They were married in New York City in 1901 and went to live in California, where their son William (Billy) was born the following year. After four years of marriage, the Warrens divorced. Little Billy, however, would spend most of his childhood on his grandmother's ranch.

Following in her sister's footsteps, Agnes set out in February 1907 to visit their Aunt Mary in Germany. She placed Norman and Loraine in the care of the Nels Field family, friends who ran a trading post about thirty-five miles north of Magdalena, and left little Morley with either Lora or Grandmother Morley. Apparently, no one saw anything strange in Agnes leaving the youngsters. Her own mother had set the example. There had always been someone "with whom to leave the children," Agnes wrote, "when [Mother] chose, as she frequently did, to accompany her husband on some interesting trip."

There is no record of how Agnes spent her days on the continent. But we do know that when she returned to the United States in the fall, her Gladstone was filled with "all sorts of lovely clothes" acquired in Paris. Before starting for Datil, however, she made a side-trip to Apalachin, New York, to meet for the first time Eugene Manlove Rhodes, then a little-known writer whose stories set in cattle country had caught Agnes's fancy.

Some years earlier, she had read several of his tales in Charles Lummis's *Out West Magazine* (later *Sunset Magazine*) and became excited by their realistic portrayal of the life she had known on the ranch. At the time, she did not connect the author with the Gene Rhodes she knew only by reputation: as the "locoed cow-puncher" of Engle, New Mexico (and the best bronco-buster in the Southwest), who carried a large scrapbook of poems around with him in his bedroll. Then one day (probably in early 1905), she received a letter from Eugene Manlove Rhodes. He had read one of her stories, found that its plot was almost identical with one of his unpublished pieces, and wanted to know how she "could take an incident so typical of our country . . . and turn it into cash." Rhodes had sold his first fiction story to Lummis for ten dollars but now hoped to receive greater "pecuniary benefit" for his efforts. Agnes replied at some length, advising him to get in touch with her editor at *Munsey's*, Bob Davis. Although Davis rejected the first story that Rhodes submitted, Davis's encouragement, as well as Cleaveland's, was just the tonic he needed to keep writing.

When Agnes met Gene for the first time, "after months of joy-bringing correspondence," he told her that his first letter to her "had been his toss of a coin: that he had been quite serious in what he had said in the letter to the effect that he felt he should probably stick to bronco-busting which was his forte but that the desire to write gnawed at his vitals and gave him no peace." If she deigned to answer, he "would turn from cow-punching to literature"; if she did not answer, he would stick to bronco-busting.

Agnes's stay at the Rhodes' residence in 1907 cemented a warm lifetime friendship with both Gene and his wife, May, who, in writing about Agnes's visit, described her as "a slight, graceful, and vibrant lady." Gene had made the acquaintance of May Davison Purple, a widow with two sons, through a letter she wrote in praise of his poetry. They corresponded for several months—May from her home in Apalachin, Gene from his ranch near Engle. They married in 1899 and shortly thereafter took up residence in Tularosa, New Mexico, where their son Alan was born in 1901. Gene went to live with May in Apalachin in 1906 and did not return to New Mexico until 1926—a twenty-year exile, in the words of W. H. Hutchinson, Rhodes's biographer. During these years, Rhodes published seven books and numerous short stories, most set in the cow country of New Mexico. He also corresponded with Agnes and, while visiting his mother in Pasadena in the early 1920s, took Agnes to visit the artist Maynard Dixon and the humorist Will Rogers. By the time of Rhodes's death in 1934, he had become "one of the most widely-read western writers" in the nation, as Roland F. Dickey noted.

Upon her return to Datil in the fall of 1907, Agnes gathered her six-year-old son and five-year-old daughter and went to live in Silver City, a mining and ranching settlement in southwest New Mexico that was known to have a good school system. To cover expenses, Agnes took over the Rosedale Dairy, about two miles outside of town. Norman rode his pony into Silver City and back each school day, accompanied on occasion by his sister Loraine, who returned home on her pony once they reached the school. For most of the time that Agnes ran the dairy, however, Loraine and the toddler Morley stayed with their Aunt Lora on the Swinging W. Norman recalled that his father, who "was doing quite well in Oroville," visited them at Christmas and other times. Still, Agnes's move to Silver City underscored her desire to be independent—to make her own tracks, as she once put it. Indeed, many years after Agnes's death Loraine indicated that her mother's flight to Silver City was in line with her thinking that women should not be so dependent upon men.

Agnes later chose to record the more humorous aspects of her stay in Silver City, as, for example, when Loraine had a run-in with the marshal for racing her pony down the town's main street. Then there was the time Agnes fired two employees who, with drawn pistols, were "getting ready to shoot it out at the barn." Thereafter, the marshal took pity on the novice dairy woman and let a prisoner out of jail to help with the milking. The most memorable incident occurred during the time she had more customers than she could supply. To have enough milk to go around, she temporarily added water to it, mixing in some butter coloring to give it the golden rich color that customers expected. One day, her employees, ex-cowpokes, failed to put in the correct amount of coloring, so that when it came out pink, townspeople thought they were being poisoned.

Agnes also befriended a young woman, Elisabeth Nichols, who taught English at the Silver City Normal School and who, in her own words, "had great ambitions toward short-story writing." After publication of *No Life for a Lady*, she recalled that Agnes had done her best to help further those ambitions. Now a teacher at MacMurray College for Women in Jacksonville, Illinois, Nichols told Agnes that her book was a required reading in her course on contemporary American prose. "There isn't a thing as good in reproducing an important era in our country's ever changing life and culture," she added.

It was in Silver City that Agnes received a letter from Gene Rhodes, with the request to help his friend Maynard Dixon "out of a jackpot." Dixon was in the early stages of a remarkable career that earned him a national reputation for his authentic portrayal of the American West on canvas. But the San Francisco fire of 1906 had destroyed his studio, including the props he used for his western scenes. Gene asked Agnes to find appropriate cowboy gear "at the least possible cost" and ship them to Dixon in New York City, where he had reestablished his studio. Agnes described in her memoirs how she filled this request—and earned Dixon's gratitude in return, as well as an inscribed painting from the artist.

In the fall of 1908, the Cleaveland children were taken to Berkeley, California (by either their Aunt Lora or Grandmother Morley), where they settled in temporary housing on Kelsey Street, with their Aunt Lora as their caretaker. Sometime later, Agnes sold the dairy and joined the children in Berkeley. The family eventually moved into a large house on Cedar Street, which, when renumbered 2532 Cedar, was to remain their permanent address for many years. Their three-story, redwood shingle home was situated halfway to the top of the Berkeley hills, close to the University of California campus, "an incredibly magnificent site," one Cleaveland youngster remembered. The hills were "covered with wild oats in the winter and tan remnants in the summer, with ravines supporting bay trees, toyon, and too much eucalyptus tending to crowd out natives. Children could play in such places." Norman and Loraine soon enrolled in the nearby Hillside School, and in due time Agnes became president of the Hillside School Mothers' Club.

Despite household responsibilities and other disruptions, Agnes Morley Cleaveland continued to work at her writing. According to the *Silver City Enterprise*, she spent part of the summer of 1909 at the Grand Canyon in Arizona gathering material for a series of stories on the U.S. Forest Service. While at the canyon, she met young Margie Gunderman, an employee of the Indian Civil Service, stationed in Cataract Canyon—a branch of the Grand Canyon. Entranced by Margie's description of Cataract Canyon and the Indian school located there, Agnes made the fifty-mile trip on horseback with Margie and stayed several days with her at Cataract Canyon.

Meanwhile, in July of that same year, *Munsey's Magazine* published Agnes's "Stoneman, Forest Assistant," the first (and evidently the last) of her Forest Service stories. This tale features a strong woman protagonist in conflict with Forest Service officials, who question her method of cutting timber. Their disagreements are resolved after she rescues Stoneman, a forest assistant, from an encroaching forest fire. The story's credibility derives from the author's acquaintance with New Mexico ranchers and their antagonism to the recently established U.S. Forest Service and its restrictions on the use of forest land.

Still later in 1909, after Agnes had returned to Berkeley, she joined with other writers to found the Press Club of Alameda County, later known as the California Writers' Club. Among its early honorary members were Jack London, John Muir, Joaquin Miller, Charles Lummis, and Ina Coolbrith, soon to be named California's poet laureate. The club held monthly dinner meetings, where one or more prominent writers talked about their work. They also hosted receptions for visiting members of the literary world. Among the most memorable was the one held in Oakland on February 22, 1910, in honor of Mrs. Frank Leslie, former head of the Frank Leslie Publishing Company. More than three hundred guests attended, the women wearing beautiful gowns and "brilliant jewels." According to one observer, "the scene was more like that at a society ball than at an affair sacred to Bohemia." Agnes, wearing a turquoise blue satin gown, was in the receiving party, along with the flamboyantly dressed Joaquin Miller, whose long curls of gray were surmounted by "a Turkish fez of red."

In 1914, the club published *West Winds, California's Book of Fiction*, the first in a series of books featuring the work of members. Edited by the novelist Herman Whitaker, this volume included London's "The Son of the Wolf," Lummis's "The Enchanted Mesa," and Cleaveland's "The Greatest of These," first printed in a 1904 issue of the *Metropolitan* magazine. Agnes remained active in the California Writers' Club until the 1940s, when she moved permanently to New Mexico.

Cleaveland eventually stopped writing for a time, discouraged by what she considered the superiority of Rhodes's literary efforts compared to her own. To help snap her out of her doldrums, Gene Rhodes encouraged her to turn a comical incident involving Corky Wallace into a short story. The cowboy, it seems, had put his brand on a "long-ear" (an unmarked calf) that most observers believed belonged to another outfit. As it turned out, however, the long-ear had been following one of Corky's own cows; so Corky had, in fact, "stolen a calf from himself." Because Gene had so enjoyed the tale, Agnes insisted that he write the story. Finally, they compromised and agreed to collaborate. But Agnes developed writer's block; and "The Prodigal Calf," published in *The Silhoutte* in 1916 with Cleaveland and Rhodes listed as coauthors, seems mainly to have been the work of Gene's pen. This frustrating experience, Agnes avowed, proved to be "the capstone of my resolution to quit that 'writing game' about which I had so smugly advised Gene some years earlier."

Still, factors other than "The Prodigal Calf" episode help to explain Cleaveland's disinclination to write any more short stories. About 1909 (or 1910), she became a Christian Scientist. For the next decade or more, the church and its teachings became—second only to her family—the central focus of her life. She also became a suffragist and seems to have taken considerable pride in the work she did to help California women gain suffrage in 1911. On a trip to New Mexico the following spring, she spoke on "Woman's Suffrage and How We Won It" before a "large audience of cowboys and miners" in Magdalena. And like her mother, Agnes actively supported the Nineteenth Amendment, which, when enough states ratified it in 1920, gave women the vote in federal elections. Later, Loraine Cleaveland would tell "wonderful stories about marching with her mother in parades in San Francisco, with a yellow ribbon across her chest reading 'Votes for Women.'"

Domestic responsibilities also increased after Agnes gave birth to her fourth child, Mary, on December 28, 1912. By this date, she and Newton probably were aware that their second daughter, Morley, suffered from behavioral or mental problems. Although no documents reveal the extent of her disability, Morley would require supervision for the rest of her life. But in the early years on Cedar Street, the family home seems to have exuded warmth and security for the children. Loraine kept a favorite pony, Topsy, on a nearby lot and made plans one year to ride in a horse show at the state fair. Norman played football during his first year at Berkeley High School, though being the smallest lad on the team he "had a hard time of it" and quit to concentrate on his studies. (He would later become a star halfback on the Stanford football team.)

In 1917 the family broke away from the city and moved to a seventy-acre farm, situated about halfway between Walnut Creek and Concord (to the east of Berkeley). Norman recalled that his father bought two cows, two horses, six pigs, a plow, mowing machine, hay rake, "and everything, for the use of his son," whose job it was to milk the cows and feed the pigs and the horses both before and after school. In their first year on the farm, Norman and Loraine rode into Concord in a surrey pulled by a horse named Kate to attend Mt. Diablo Union High School. In their second year, they arrived at school in the family's new Model T Ford. Meanwhile, during this two-year interlude, Newton commuted to his office in San Francisco, taking a train to Oakland and then a ferry across the bay to the city. We can only speculate how Agnes spent her days on the farm—tending to the needs of her children, keeping abreast of news about World War I and the suffrage movement, and continuing her study of Christian Science. And most assuredly, she would have helped manage the farm and taught her oldest children how to operate the farm machinery.

During the 1910s and 1920s, Agnes made regular trips to the Datils, where she quickly reentered the life of the ranch. She helped round up cattle, tear down barbed wire fences (which the Forest Service ordered them to do), and went for supplies in Magdalena (now in an automobile rather than in a horse-drawn wagon). Sometimes Agnes's stay coincided with visits from Dane and Mary Roberts Coolidge, friends "whose advent was always hailed with delight and whose stay was always too short."

The Coolidges made their first trip to the Datils in 1911, the same year Dane published *The Texican*, his second western novel. Mary later avowed that from 1911 on, Dane "was the devoted friend of Ray Morley." During his lifetime (1873–1940), Dane Coolidge published more than thirty-five westerns, as well as one hundred or more short stories in magazines. At least two of his novels were situated on or near the Morley ranch. And in appreciation for Ray Morley's hospitality, Dane often sent him autographed copies of his books.

While Dane Coolidge and Ray Morley traveled together about the countryside, Mary often spent time with Agnes's mother, whom she described in a letter to Agnes as "the only intellectual woman I knew in that vast region." A talented scholar and writer, Mary Coolidge published in 1912 *Why Women Are So*, now considered a classic in women's studies. Prior to Ada Morley's death in 1917, Mary wrote an essay entitled "Ada McPherson Morley, Non-Conformist," the title suggesting the author's admiration for the older woman. This treatise became a chapter in Coolidge's unpublished manuscript, "They Carried Live Coals," a book about unusual women who were "ahead" of their time.

Ada Morley was indeed a nonconformist, in the sense that she supported causes that the American people had not fully embraced. Nor did she always conform to the expectations of her family and neighbors. In the last decade of her life, her eyesight steadily deteriorated until she was nearly blind. Determined to manage her own affairs and not be a burden on her children, she moved into a cottage near the ranch house and was assisted by a succession of sympathetic ranch hands, who made her fires, cooked, washed dishes, and read to her. Casey Jones, who worked in this capacity for two years, drove her to California in 1915 in a Model T Ford to see her daughters and their children and to attend the Panama-Pacific Exposition in San Francisco.

Agnes and Lora later persuaded their mother to remain on the West Coast, alternating her time between the Cleavelands in Berkeley and Lora in Palo Alto. Before long, however, Ada became restive and took off for New Mexico—without telling anyone—using public transportation to reach Albuquerque. When Agnes went to see her mother the following year, Ada's "unofficial caretaker" was an eighteen-year-old cowboy named Lyle Vincent.

Despite her loss of vision, Ada Morley worked assiduously for women's suffrage, a cause that had absorbed her interest for more than thirty-five years. Dictating to her assistant, she kept up a steady correspondence with both national and state suffragists, urging political action to gain votes for women. With New Mexico being the only western state without suffrage, Ada felt keenly her own lack of political citizenship. In a letter dated February 24, 1916, to Anne Martin of the Congressional Union (a national suffrage organization), she expressed her outrage: "I'm always and ever on the alert to gain my own liberty. Disfranchisement is a disgrace."

Unfortunately, Ada fell seriously ill with kidney trouble in the late summer of 1916; consequently, her daughter Lora returned to Datil to care for her. Convinced that her mother needed contact with other people (since she was not bedridden), Lora leased the Baldwin place, with its store, hotel, and post office, and took the civil service examination to become the Datil postmistress. With tourists and locals coming by daily, Ada would not lack for companionship. Sometime after the start of the new year, however, Ada moved into her son's cottage in Magdalena. Although she visited her daughter in Datil in the spring, she became ill during the winter and died at her home in Magdalena on December 9, 1917, at the age of sixty-five. Two funerals were held, one in Magdalena and the other in Las Vegas, where she was buried beside the remains of her husband, William R. Morley. The Las Vegas press eulogized Ada in these words: "Mrs. Ada Morley, author, philanthropist, the old-time cattle queen of New Mexico, whose residence in the state antedates the railroad, was one of the intellectual aristocrats of the southwest whose passing is a loss to the state." It is presumed that Ada's three children attended the graveside services. Days later, a notice appeared in the *Magdalena News*, signed by Agnes, Ray, and Lora, extending their "heartfelt thanks for the kind assistance and sympathy extended us during the illness and death of our beloved mother, Ada M. Morley."

At the time of Ada's death, Ray Morley was serving a two-year term (1916–18) as president of the New Mexico Cattle and Horse Growers' Association. In *No Life for a Lady*, Agnes recorded that these were good years for New Mexico stockmen. The outbreak of World War I had created an increasing demand for American beef in Europe; the home market also expanded, especially after the United States entered the war in April 1917. From then on, the federal government encouraged cattlemen to raise more beef as a patriotic duty. "By 1918 the demand for beef swelled to unprecedented proportions," one historian writes. "Cattlemen sold all they could, raised all they could, and got the highest prices they could. Prices shot up to the highest levels in history." Like other ranchers, Ray Morley stocked his ranges to full capacity.

Morley's major problem during the war years was a shortage of ranch hands, since almost all of his employees either were drafted into the army or volunteered. Consequently, Ray relied on Navajo Indians to manage his flocks of sheep. And in the summers of 1917 and 1918, he enlisted teenage boys home on vacation or visiting relatives to ride herd on his cattle. Agnes's son, Norman, worked for Ray both summers and has left an engaging account of his tussles with cattle and with his Uncle Ray in *The Morleys, Young Upstarts on the Southwest Frontier*. The teenagers were known affectionately as Morley's "Kindergarten Outfit," captured on film by Dane Coolidge during his stay at the Drag A Ranch in 1918. Lora's son Billy also was part of the crew, as were Charley and Willie Anderson, Ira Wyche, and Langford Johnston. Tom Reynolds, the Drag A's foreman and chief of the Kindergarten Outfit, married Agnes's sister Lora during the summer of 1918; the couple would spend the rest of their lives in the Datil area.

At war's end, large numbers of Americans set out in their automobiles to see the West, a phenomenon that did not go unnoticed by either the *Magdalena News* or the Morley siblings. On October 16, 1919, the press reported: "The tourist travel has been exceedingly heavy for the past two weeks, scores of cars passing through Magdalena every day, and at night the hotels and camp yards are filled with the travelers. . . . By six o'clock [Sunday night] one hundred cars had arrived in town and a number more coming in later. It was estimated that over three hundred tourist travelers passed the night in Magdalena."

To take advantage of the tourist trade, Ray—with Agnes and Lora's blessings—dismantled the White House in the fall of 1919 and moved it log by log to Datil, where it was reassembled, enlarged, and renamed "The Navajo Lodge." Completed in 1920, the new hotel was a scrumptious affair. A newsman wrote this description for his readers:

> The lodge is built entirely of hand hewn pines from the Datil National Forest and has a large attractive lobby reaching clear up to the roof with rustic stairways on each side leading up to a mezzanine floor from which the bed rooms open. A large open fireplace adds cheerfulness to the lobby. The ceiling of the lobby is supported by logs thirty-five feet long, which are covered with hewn slabs laid herring bone style. On the ground floor leading off the lobby is a dining room that will accommodate twenty-six guests comfortably, with a modern equipped kitchen adjoining.

The lobby and dining room were decorated with large Navajo rugs and hides of bears, mountain lions, wolves, coyotes, wildcats, and other predatory animals killed in the Datil Mountains. The building had electric lights and "hot and cold water and modern plumbing" in each bedroom. Ray also built a number of hogans on the premises and enticed some Navajo families from the lower end of Alamosa Creek to live there.

(The men often found jobs on the Morley ranch, while the women wove blankets to sell to tourists.)

Much to the delight of Ray, Agnes, Lora, and other Datilites, the Navajo Lodge became a fashionable landmark for the entire region. Political and lodge meetings were held there, as well as wedding celebrations, Fourth of July dances, and sundry other merry-making gatherings. Hunters and tourists alike enjoyed evenings spent in front of the fireplace listening to Ray Morley tell "stories of bygone days."

But, while the lodge did a booming business, Ray Morley's ranching operations fell on hard times. Soon after the armistice, "the cattle boom collapsed." Cattle prices dropped precipitously, drought seared the ranges, and stockmen struggled to repay government loans that had been granted as a wartime emergency. Charles M. O'Donel, manager of the Bell Ranch in northeastern New Mexico, described the situation in 1921: "The country is in a deplorable condition from a business point of view. Nobody is selling cattle or sheep; nobody has either money or credit to buy anything. . . . One hears on all hands of stockmen who were carrying loans on their property taking bankruptcy."

Agnes returned to the Datils in the fall of 1921 and witnessed Ray's fight to hold on to his ranch. She was accompanied by at least one of her daughters and by George McManus, a family friend and a concert pianist of some note who gave piano lessons to Mary Cleaveland. Something of a child prodigy, Mary would later recall that "George McManus was always in and out [of our Berkeley home] and he had tried to give me piano lessons earlier than I can remember him. . . . he was as much as a member of the family."

Starting in November 1921, Agnes took over management of the Navajo Lodge, thus freeing Ray from some of his administrative chores. There seems to have been a special bond between brother and sister, probably forged during those early years of trying to help manage their mother's ranch. Mary Roberts Coolidge once remarked to Agnes: "[Ray often] told of your youthful escapades together—roping, riding, chasing wild game in the mountains. Always he included you with [the] cowboys." And Laurence Lee, who knew Ray about all of his life and was closely associated with him during the difficulties in the 1920s, wrote in a letter to Agnes after he had read *No Life for a Lady* that Ray never had "what might be called close or intimate friends. I am sure that you were the most 'intimate friend' that he had and were no doubt more in his confidence than anyone else."

So while Ray met with other stockmen to address their mutual problems, Agnes oversaw operations of the lodge, presiding at special dinners (among other duties) and changing beds and cooking for tourists when her employees left with little warning. She had bought a new Model T Ford in Magdalena (where Ray held the Ford dealership)—to be "absolutely independent in going and coming," she said. It soon became a local joke, however, that because she loaned the car to so many of her friends, she often did not know its whereabouts.

By early 1922, both Mary and Morley Cleaveland were living with their mother and attending school in Datil's one-room schoolhouse, where cousin Billy Warren was their teacher. Later that summer, Agnes met her oldest daughter, Loraine, in Magdalena and then drove her to Datil in the Ford. Typical for travelers in those days, they had two flat tires and then engine trouble five miles from their destination and "had to send for help to get in."

Meanwhile, Ray Morley had devised a plan to move his cattle and those of his neighbors from the drought-stricken ranges of New Mexico to the Mexican state of Chihuahua, where grass grew in abundance (due to the Mexican Revolution that had decimated the cattle herds). Morley toured Chihuahua in September 1922, negotiated with Mexican authorities, and then traveled to Washington to secure approval for his plan. He then completed negotiations on a return trip to Mexico.

Ray's Drag A outfit started gathering and branding steers in October for shipment across the border. Other stockmen were shipping their animals outside the state as well. In early November, the F. A. Hubbell Company filled sixty-seven railway cars with sheep that were destined for Colorado—"the longest train that has ever left Magdalena," declared the *Magdalena News*. Two weeks later, Ray Morley and Theodore Gatlin shipped 2,165 cattle to Mexico, and more shipments were to follow.

Agnes witnessed this "epic hegira," as she called it, "those slow-moving skeleton-like herds that were prodded along to the railroad-shipping points and loaded into cattle cars for a destination that too many of them were never to reach. Carcasses were hauled from the cars whenever the train stopped." She was in Magdalena the night that Ray shipped the last of his steers to Mexico and ran errands in her Model T—"hauling saddles and bedrolls from stockyards to railroad station, hauling cowboys from stockyards to town, or the other way around."

Agnes returned to Berkeley in the late summer of 1923, only a few weeks before a catastrophic fire destroyed the family home on September 17. Early that afternoon, a brush fire started in Wildcat Canyon (today's Tilden Park) and, "fueled by a strong dry northeasterly wind," the flames roared over the crest of a ridge and swept into a residential district. The fire spread from house to house, sparks igniting the redwood shingle roofs. When the wind subsided shortly after 7:00 P.M., firefighters checked the advance of the fire, but nearly six hundred homes and buildings had been destroyed and four thousand people left homeless. Fortunately, hundreds of university students had kept the flames from advancing onto the grounds of the University of California.

Agnes recalled that she had been given exactly two minutes to get out of the house "after it was certain that it would burn." All of the important family documents—her mother and father's diaries, personal letters, business papers, manuscripts—were left behind. Loraine and another daughter struggled to get the family car through streets clogged with fire watchers and people trying to move household goods out of the area. Loraine remembered that when she finally met up with her mother she was wearing "three large hats, an Alaskan sable cape and carrying two oil paintings slung around her neck." Agnes had saved the Maynard Dixon painting.

That night soldiers and National Guardsmen cordoned off the burned area to prevent looting. The next day Agnes and other residents who had lost their homes were permitted to return and search among the ruins. She found little of value. "Well, there was the bathtub, looking a little surprised and not a little triumphant, sitting in the middle of the concrete square that had been the furnace room floor. In its two story drop it had landed right side up and it alone of a ten room complement of home furnishings had survived the fire in recognizable form. The piano was a mass of tangled wires, the kitchen stove a blob of tortured metal."

Life soon resumed a degree of normalcy, however. The Cleavelands took up temporary residence on Ashby Avenue, south of the university, and later moved into a new house constructed in a pueblo style on the old site. About a year after the fire, Agnes returned to Datil, and presently her nephew Billy Warren reported to Norman Cleaveland that she was "getting along fine and seems to be enjoying her visit very much. Uncle Raymond uses her car often though, so she has a time keeping up with it."

Christian Science also remained central to Agnes's life, even though she recently had been excommunicated from the church. To understand how this stricture came about, it is necessary to reconstruct her early days as a Christian Scientist. About the time she returned from Germany, Agnes underwent a period of spiritual searching. She found little comfort in the dogmas of orthodox Christianity so when a friend handed her a copy of Mary Baker Eddy's *Science and Health* she read the entire six hundred pages seeking enlightenment. It came slowly. Finally, she joined the church (around 1909 or 1910); shortly thereafter, she met the charismatic Herbert W. Eustace of San José, with whom she took "class" to more fully understand the tenants of Christian Science.

She entered wholeheartedly into the affairs of Berkeley's Fourth Church of Christ, Scientist, on Fulton Street, holding various offices, including reader and then president. Mary Cleaveland remembered having been taken to church on

Sundays and some Wednesdays. She also recalled that a congenial group of church members met at each other's homes to talk about Christian Science. A family friend would later write to Mary, "The group that sat at [Agnes's] feet and listened enthralled to her tales were interested in Christian Science, but that isn't all that held them together. We all loved your mother and were intrigued by her thrilling adventures and her sense of fun."

In due time, Agnes became a church-authorized practitioner, that is, a healer, after completing a course of study. She had joined the church for spiritual reasons, but for many others the great drawing card was physical healing: they became church members after they, a relative, or a good friend "experienced a physical healing through Christian Science." Moreover, for Agnes and many other Christian Science women, becoming a practitioner provided a way "to engage in public service, attain self-sufficiency, and prosper." Like other healers, Agnes maintained a public office and probably charged fees equal to those of local physicians (as decreed by Mary Baker Eddy). Grateful patients wrote to her, settling their accounts, and thanking her (as one woman put it) for "your very patient and loving care of me." Although none of Agnes's children seems to have become a Christian Scientist, they did respect her religious beliefs. Gene Rhodes likewise accepted her choice of religions, though in 1915, with tongue-in-cheek, he wrote to a friend: "Mrs. Cleaveland was once a very brilliant lady but is now a Christian Scientist. (Joke—Persiflage) She is my very good friend and she may be a Buddhist if she likes unmolested by me."

For more than a decade, Agnes considered herself a devoted and obedient member of the church. Two of her essays, "What Is a Christian?" and "Autocracy, Theocracy, and Democracy," appeared in Christian Science journals. Then in 1918 she was shocked to read in a newspaper that the Board of Trustees of the Christian Science Publishing Society had filed suit to obtain an injunction against interference in their administrative affairs by the Board of Directors of the Mother Church (both boards were located in Boston). What followed was a long, drawn-out court battle, known as the "Great Litigation"—a struggle for power that badly damaged the morale of many Christian Scientists.

Doubly shocking to Agnes was the fact that Herbert W. Eustace, her former teacher and mentor, was one of the two trustees who had sought the injunction. The case was finally settled in 1921, when the Massachusetts Supreme Judicial Court ruled in favor of the Board of Directors. Thereafter, the board excommunicated Eustace, though he continued to be active on the West Coast, teaching large and frequent classes and publishing works on Christian Science. Many others who disagreed with the court's decision resigned from the church and became, in the words of Charles S. Braden, "free-lance Christian Scientists." Agnes sided with her teacher and was also excommunicated. Later she admitted that her expulsion "was the result of extreme provocation. I forced the hand of the church authorities by telling them in the most [powerful] language of which I have command that I considered them wholly unfit to be church officials." Thereafter, she too became a free-lance Christian Scientist, adhering to Christian Science principles as best she could and ministering to those who sought her help.

In the early twentieth century, Agnes attempted to balance her family life with an increasingly busy agenda as Christian Scientist, clubwoman, and political activist. In both her public and private lives, she found joy and personal fulfillment—as well as conflict and disappointment. One of the greatest pleasures of her middle years, however, was reentering the life of Lou Henry Hoover, former Stanford schoolmate and first lady of the United States. Through it all, the good times and bad, Agnes approached life in her own inimitable style—with courage, passion, humor, and a good measure of rebelliousness.

CHAPTER 4

"With Sleeves Rolled Up[,] Breathing Fire and Brimstone"

ON December 28, 1937, Agnes Morley Cleaveland wrote to Lou Henry Hoover thanking her for the raisins that she (Lou Henry) had sent to the Cleavelands. Earlier in the decade, Agnes had reestablished her friendship with Lou Henry and thereafter entered the social and political world of the Hoovers. In this 1937 letter, Agnes expressed her belief that when reason returned to the country the name of ex-President Hoover "will be emblazoned in even more glowing letters," alluding to the fact that many people wrongly blamed him for the Great Depression. She also signified her willingness to work on Hoover's behalf. "I await any call to service," she wrote, "whether it be to 'stand and wait' or go forth with sleeves rolled up breathing fire and brimstone. The latter role better suits my inclination but I can manage the self-restraint for the former."

This well-turned phrase, "with sleeves rolled up breathing fire and brimstone," captures perfectly Cleaveland's approach to the career she carved out for herself as clubwoman and political activist during the first decades of the twentieth century. These were years when middle-class American women exuberantly embraced club work, finding outlets for their talents and energies through a multitude of organizations and civic reform projects. Agnes became a leader in every organization she joined and thus she fits historian Anne Firor Scott's description of early twentieth-century heads of women's clubs: they were ambitious, well-educated, enjoyed public speaking, and "were not on the whole, given to modesty."

Agnes was most actively involved in civic and political affairs during the 1930s. Still, she had made herself known in the Bay

Area as early as 1909, when she became a founding member of the California Writers' Club. And, soon after women gained suffrage in California (1911), she helped found the Berkeley Civic League, forerunner of the Berkeley League of Women Voters.

Almost nothing is known of Cleaveland's political activities in the 1920s. We do catch glimpses of her social and family life, however, in newspaper clippings and in a few surviving letters. Foremost on her social calendar were the meetings of the California Writers' Club. Possibly it was the camaraderie of fellow members that inspired Agnes to continue writing, despite her disclaimer following publication of "The Prodigal Calf" (1916).

Agnes was one of twenty California writers (most, if not all, members of the California Writers' Club) who contributed to the serial novel *The Trail of the Serpent*, published in the *Oakland Tribune*, starting in the spring of 1922. Two years later her short story "The Drawn Line" appeared in the *Overland Monthly*. Like many of her previous tales, "The Drawn Line" was based on a childhood experience that would reappear in *No Life for a Lady*. It also is included in the club's third volume of *West Winds* (1931), an anthology of members' recent publications.

As happened so many times in her career as a clubwoman, Agnes became a driving force in the California Writers' Club. She was elected to its board of directors in 1927 and served as its president ten years later. Her good friends Mary Roberts and Dane Coolidge also were active in the club, Mary having been elected as vice-president in 1924 and both frequently speaking at the dinner meetings. Agnes penned a profile of Dane for the December 1929 issue of the *Overland Monthly*, an issue totally devoted to the works of California writers. Her short essay, "Three Musketeers of Southwestern Fiction," featured two other writers as well, Eugene Manlove Rhodes and Oklahoma-born John M. Oskison.

During the twenties, Agnes continued to minister to Christian Scientists, to attend functions of the Alpha Phi sorority, and to make regular treks to the Datils. Newton seems to have been making a good living at this time, his earnings as a mining engineer sufficient to maintain a large house and to employ gardeners and household help. Then, too, Agnes's remuneration for her short stories and Christian Science work would add to the family's income and help pay for her trips to New Mexico.

Throughout the decade, Agnes enjoyed exchanging visits with the family of Newton's older brother, William Cleaveland, who prior to the Berkeley fire lived near Agnes and Newton on Cedar Street. Later, William and his family moved to Benicia about twenty-five miles farther north, close enough for the two families to socialize. That Lois Cleaveland, William's youngest daughter, chose Agnes and her daughter Loraine as two of her four bridesmaids in 1924 suggests the closeness of the two families.

In contrast, a series of letters that Lora Reynolds wrote to Agnes in the 1920s reveals something of the complex and oftentimes difficult relationships that existed between Lora and her two siblings. From childhood on, the three Morleys exhibited similar characteristics; they were high-spirited, strong-willed, independent-minded, and often outspoken. That clashes occurred among siblings with such strong egos should not be surprising. But *No Life for a Lady* fails to prepare the reader for the sharp-tongued remarks found in Lora's correspondence. Certainly part of her pique can be attributed to the fact that the Cleavelands and Ray Morley had more money than the Reynoldses, but only part. In one outburst, Lora asked Agnes, "Did it ever occur to you that your Ego was entirely too great for anything you have ever accomplished in reality? Since birth you have dominated someone, if it had to be by howling your head off to get your way. . . . Both you and Ray have grown so in your own estimation that no one else is allowed a mind." Later she wrote of Ray: "He has treated me very badly, and I do not feel that he has shown a noble trait in his relationship for the last 30 years."

Although none of Agnes's letters to her sister have survived, it is not likely that she accepted Lora's chastisement without comment. Agnes could be equally as blunt and outspoken as Lora, but Lora seems to have had the more contentious personality.

Agnes's frustration in dealing with her sister's behavior is expressed in a letter she wrote to her daughter Loraine: "[Lora's] weird capacity to create trouble almost has my goat. She strikes behind your back. To your face she is usually fair—until cornered—and you are misled into feeling that you can deal with her. I often feel that I have misjudged her. She seems so rational and kindly disposed and then wham! She has hit a body blow when I wasn't looking."

But families do stick together despite arguments and harsh words. Shortly after Lora's diatribe against Ray, she sent a plea to Agnes to "get in behind the move to have Ray appointed to fill Sen. A. A. Jones [recently deceased] place in the U. S. Senate. Ray has every thing to make him a success as U. S. Senator and he can serve the state with his hands free. He has no personal interests to make people feel that he will graft." Lora also appreciated some of the little things Agnes did for her—sending articles of clothing and kitchen accouterments, for example, and extending loans to members of her family. Then there were the larger gifts, such as helping to finance Lora's trip to Hawaii in the summer of 1929 to visit their Stanford friend Emily Dole, the niece of Sanford B. Dole, the first governor of the territory of Hawaii. During Lora's absence, Agnes took over management of the store and guest cottages that Lora and her husband had recently built in Datil.

The year 1929, in fact, was in many ways a remarkable one for the Morleys. Lora had her trip to Hawaii, Agnes minded the store, and Ray marched in President Herbert Hoover's inaugural parade, which led to his unexpected reunion with a son he hadn't seen since infancy.

Morley had gone to Washington to persuade Congress to award southwestern cattlemen the 17.5 million dollars he claimed they had lost through the War Finance Corporation during the postwar years. Though nothing came of his efforts, he did gain wide publicity from his appearance in Hoover's parade. His ten-gallon hat, shoulder-length hair, and flaming red beard made him a striking figure among the conservatively dressed dignitaries.

The account of how Ray and his son were reunited in Washington "reads like a fairy story" (as one reporter put it). While at Columbia University (where he was known as Big Bill), Morley had married "a girl with piercing black eyes." "Then, shortly after the birth of William Raymond Morley, [the] second, in 1903, Big Bill, for reasons known only to himself and his wife, left for the West. Mrs. Bill never saw him again." And so the story advances to 1929 when a relative sent a newspaper clipping to young Morley, who was living in Carbondale, Pennsylvania, with his wife, Margaret, and baby son, William Raymond Morley III. The clipping stated that William R. Morley, a New Mexico cattleman, was staying at the Mayflower Hotel in Washington. After young Bill made further inquiries into the elder Morley's identity, the son and father arranged to meet in the capital. That summer Bill and his family, which now included a new baby girl, spent the summer in Datil. But the senior Morley's declining health left little time to reenter the life of a son he had never known.

Ray Morley had developed serious heart problems in the late twenties. After doctors advised him to live at a lower altitude, he sold his New Mexico ranches and, in 1930, moved to Pasadena, California, where his wife, Nan, resided much of the time. On a return trip to New Mexico that summer to handle legal matters, he suffered a heart attack. In *No Life for a Lady*, Agnes described one of her last meetings with her brother. "He was leaning back in a down-stuffed chair in a Southern California hotel," she wrote. "Over-stuffed chairs after the hard leather of a saddle were to him as insufferable as air to a fish. 'Bogging down in luxury is a horrible death,' he said prophetically." He died following a severe heart attack at his home on May 27, 1932—at the age of fifty-six.

Lora Reynolds flew to California to attend her brother's funeral. Presumably, Agnes was in attendance also. Then in June she set out from Berkeley on her annual visit to Datil, motoring across country with her daughter Mary and family friend Helen Whitney. Their Ford sedan, heavily ladened with baggage, including sleeping bags tied to the running boards, frequently boiled over and suffered other mishaps on the one thousand-mile journey. Agnes would call it "the worst auto trip of my career." She documented her ordeal in letters written to the two N's—Newton and Norman—letters that easily could have been turned into a humorous piece for a popular magazine. These epistles also stand as testament to Agnes's pluck and resiliency.

"Off to a bum start," she wrote of their first day of travel. The temperature had reached 112 degrees by midmorning, and "we boiled, meaning the car as well as ourselves." They stopped in Modesto, California, to get water "for the umpteenth time," and then the car wouldn't start. Three hours later, after a local mechanic located and fixed the problem, they drove on and "boiled and boiled and boiled," stopping at every service station along the way for water. They ate supper in Bakersfield, went over the grade at Tehachapi in second gear at about twenty miles per hour, and reached Mohave at 1:00 A.M., where they spent the rest of the night in an auto camp. "Hot wind all night and noisy trains. Not too restful," Agnes wrote.

The next day they stopped at every filling station on the route to Barstow and there jettisoned some of their luggage, shipping it on ahead to lighten the load. With the car still boiling, they went on to Newberry, where they met up with Nan Morley and her daughter, Faith, who had driven over from Pasadena to join Agnes on the final leg of the journey. "It's cooler this afternoon, only 104 [degrees]," Agnes noted in jest. They took off again at 6:00 P.M., hoping to avoid some of the desert heat. But they boiled "harder than ever and just made it from one filling station to another by using the desert bag full of water between stations." They reached Needles at 11:00 P.M., took time to eat, and then drove to Oatman, Arizona, where at 1:00 A.M. (with the car "spouting like a geyser") they rousted the proprietor of an auto camp out of bed to rent cabins. Though they planned to sleep late on the third day of their travels, Agnes awoke at 5:30 A.M., "feeling that I was in a furnace"—and so they went on, only to break down again in Kingman.

They finally reached Holbrook, Arizona, at 9:30 P.M. Although totally exhausted, Agnes now had to locate Nan, who had gone on ahead with all of the sleeping bags while Agnes's car was being fixed. Agnes found her sister-in-law only "after I had started to walk to their auto camp and had fallen into a deep ditch along side of the road in the dark. I skinned both knees and did worse to my temper." The caravan had no especial trouble after leaving Holbrook and reached Datil at 6:30 P.M. at the end of the fourth day.

A few days after their arrival, Nan and Faith Morley, accompanied by Lora and Tom Reynolds, scattered Ray Morley's ashes, some at the site of the old White House and some on the top of the divide that separated the Datil Mountains from the Alamosa Creek region to the north. Agnes apparently was attending to business elsewhere and missed the ceremony. A letter to the two N's in early July, however, described her latest misadventure. "The jinx still rides on our shoulder," she wrote. "Mary and I started in to town to get the small freight and the things shipped from Barstow. Six miles from Magdalena the car caught fire and was blazing under the hood before we got our wits together to know what was happening. We threw dirt on the blaze and after quite a battle got it out." Milt Craig, Ray's former partner in the Ford dealership, received word of their predicament and drove out to push them into town.

Agnes's problems were by no means over. Midway through her stay in Datil, she broke a tooth, which caused so much pain that she was unable to eat. It became necessary to pull seven of her lower teeth and replace them with a plate. "Am going to try it without any anesthetics, local or general," she vowed. We know that her dentistry bill amounted to seventy-five dollars; we do not know if she kept her vow.

Foremost in Agnes's mind, however, was the welfare of her twenty-seven-year-old daughter Morley. For at least the past two years, Agnes had employed Mrs. Mary Thomas as Morley's governess or caretaker. Agnes had spent the summer of 1931 in Datil looking over land near the ranch of Mary's son, John Thomas, who hoped to get financial backing from the Cleavelands to expand his holdings. At the same time, Agnes wanted to acquire property on which to build a house for Morley and

Mrs. Thomas (both of whom were spending the summer with John and his wife, Sayward). This scheme, Agnes reported to Newton, "satisfies Morley. She seems more contented than I've ever known her. She has not had a tantrum since she was here. She has her horse, Prince, and I believe if she had ten head of cattle shd [*sic*] take on a stability and interest that would go a long way toward inducing normality."

Agnes continued to pursue this plan the summer of 1932. Morley and Mrs. Thomas had preceded her to Datil and were staying with the John Thomas family. For part of the summer Agnes and Mary pitched their tents on the Thomas property. In a letter to Newton and Norman, Agnes complained of the "high wind, that parches you through and through. Tents flap all night and are uninhabitable by day on account of heat." Half in jest, she concluded: "If the wind would stop blowing and it would rain, I'd feel better. I *don't love anybody*. So there. Mom." But nothing came of her plans to build near the Thomases. When Agnes and Mary left for California early in August, Morley stayed behind with Mrs. Thomas to spend the rest of the year on John and Sayward's ranch.

Upon her return to Berkeley, Agnes plunged into her duties as newly elected president of the Political Science Club, a nonpartisan group of women dedicated to the study of government. Women's clubs had enjoyed something of a golden age at the turn of the century, when memberships seem to have reached all-time highs. Although clubs had declined during the 1920s, they rebounded during the Great Depression. For many middle-class women like Agnes, club activities became a significant component of their daily lives. Not only did clubs provide companionship, they also gave women the opportunity to debate and speak publicly, to study political issues and social problems, and to work for civic improvement.

Cleaveland would preside at the Political Science Club's twice-monthly meetings, held on the second and fourth Fridays. The first meeting was devoted to a program (usually a speaker) and a business session; the second was a seminar, at which the women discussed a variety of topics, including issues relating to upcoming elections. In past years, the women had invited such notable guest lecturers as the anthropologist Robert Lowie, who spoke on the origins of the state, and historian Herbert Eugene Bolton, who discussed Spanish laws that affected women. Norman Cleaveland also spoke to the club, in January 1933, about his recent experiences in Malaya as a mining engineer.

And Agnes seemed always ready to take to the podium. At a February meeting, she debated San Francisco attorney Matthew Tobriner on the timely topic, "Technocracy as a Solution to the World's Economic Problems." In California circles, the idea that engineers and technicians "should run the economy along scientific grounds" was being discussed vigorously. Among those who admired this concept was Upton Sinclair, the 1934 Democratic candidate for governor of California. Agnes, who would actively campaign against Sinclair, spoke for the negative; Tobriner, who went on to become a justice of the California Supreme Court, spoke for the affirmative.

Clearly, Agnes took pride in her work with the Political Science Club. Upon being reelected to the presidency for the third straight year, she wrote to Lou Henry Hoover in April 1934, inviting her to attend a reception and installation ceremony, saying that Lou's presence would "make the occasion memorable." "Aside from the personal gratification," she added, "there is the angle that the Political Science Club has some of the staunchest and most understanding Americans to be found anywhere. Some are Republicans and some are *bona fide* Democrats, but all are splendid women." This statement, besides confirming the Cleaveland-Hoover friendship, reflects Agnes's penchant for nonpartisan action to reform national and local government.

While juggling Political Science Club agendas, Agnes found time to address church groups and other women's clubs on a variety of topics: American Motherhood, Is Christianity Threatened in America? What It Means to be an American Citizen, Famous Women Leaders of the 90s. But her career as a public lecturer was truly launched following the publicity she received for refusing to take part in a Franklin D. Roosevelt Birthday Ball.

Staged nationwide on January 30, 1934 (the president's birthday), the Roosevelt Birthday Balls were touted as a means of raising money to fight infantile paralysis. East Bay communities jointly organized a birthday bash that was expected to draw more than ten thousand citizens. When Agnes read in the *Berkeley Gazette* that the mayor of Berkeley had appointed her to the planning committee, she declined to serve. She explained her reasons in a letter to the editor, which was published on January 18. "I opposed Mr. Roosevelt's election, the methods by which he was elected and I continue to oppose his basic policies." Furthermore, she wrote, "this birthday party, an innovation in America for a living president, while launched ostensibly to further a philanthropic enterprise, is, as every one knows, but another instance of the ballyhoo and propaganda designed to sweep the American people along a political road entirely foreign to their traditions and the principles upon which the nation was founded." Agnes's letter struck a chord with readers who feared that Roosevelt's New Deal would destroy traditional American values. After its publication in the *Gazette*, she received scores of invitations to speak before women's clubs and patriotic organizations.

On February 4, Agnes and Newton spent an enjoyable evening as dinner guests of Lou Henry and Herbert Hoover at their residence on the Stanford campus. Without doubt, the conversation turned to politics and Agnes's letter to the editor. The Hoovers sent her home with two volumes of the former president's speeches. Days later, Agnes wrote to thank them for their hospitality. She had read Hoover's speeches and reported: "They are the real voice of America and somehow or other that voice must be heard again. Couldn't there be some sort of forum or seminar for training of a select group who could speak and write and educate the public?" She had never been part of the Republican Party organization, she confessed. "I'm just a hot-under-the-collar American and propose to say so. But since I stuck my tongue out at the Roosevelt Ball racket, I'm being asked by many people: What can we do?" Her answer: "Yell and keep on yelling."

During Hoover's presidency (1929–33), Agnes had on occasion publicly defended his policies. He espoused a "philosophy of traditional Americanism" that was dear to her heart. After Hoover left office, he delayed a year or more before he attacked the New Deal publicly. He told one associate, "[I] resolved when I left the White House that no word of mine, nor any act, should weigh in even the slightest degree, directly or indirectly, against recovery." Privately, he expressed his belief that the Roosevelt administration was marching the country into a dictatorship (a belief that Agnes shared), and he encouraged others to attack the New Deal openly. In following months, Agnes became one of Roosevelt's most vocal critics and an ardent defender of Hoover's record.

The questions Agnes posed to Lou Henry about obtaining copies of her husband's *American Individualism* (1922) led to their joint crusade to educate college students and other citizens about the American system of government. In a letter addressed to "My dear Agnes Morley," dated March 20, 1934, Lou Henry reported on the results of her inquiries. *American Individualism* was not available on or near the Stanford campus nor in the San Francisco bookstores. Nor was it on Stanford reading lists for citizenship courses. Instead, the lists were "top-heavy with outright socialistic stuff." She asked Agnes to help compile "a list of books and articles which we would recommend [to] college readers and friends. Not of a partisan political nature, but dealing with national and human affairs." She wanted people to be exposed to the "right literature."

Later in May, after Lou Henry had spent part of two days talking to college students at Whittier and Scripps, she dashed off a letter to Agnes suggesting that she (Agnes) speak to the students on the American tradition before the spring term ended. "I could not bear the thought of their scattering for the summer to so many towns and villages without a message to keep them and their family and friends thinking along constructive lines during the summer." Although final exams prevented scheduling any talks just then, Agnes agreed to speak to college students in the fall.

Lou Henry's contact with students convinced her that citizenship courses needed to be overhauled. She told Agnes of the conversation the Hoovers had with two young men who were about to graduate from Stanford. They had only "the haziest ideas of what was going on in our country or of what its foundations are." They seemed "pretty well grounded in the general outlines" of communism, socialism, fascism and Hitlerism, but when "my husband asked them what they thought of the American system in comparison, they looked very blank and one said, 'Why, we haven't got a system, have we?' And the other one rather giggled and said, 'No, haven't we just grown up like Topsy?'" They knew nothing of the Bill of Rights or of the principles of the Founding Fathers. Probably half the graduating class, she stated, must be like these "two perfectly typical intelligent, more than average American boys." More troublesome, a small percentage "have imbibed so much from the teaching of Communism and Socialism that they get here with no antidote that they are going out with our label on them, absolutely Red."

Over the following two years, Agnes gave careful thought to what should go into such a course as Hoover envisioned. Among her unpublished manuscripts is one of ten pages entitled "Suggested Outline for a Course in American Citizenship," in which she expressed her ideas on the Constitution, the two-party system, democracy, and the capitalistic system. A longer manuscript, "America: My Country," presents some of the same ideas in eleven chapters, each three to four pages in length. The most successful of her endeavors was the *American Primer*, a twenty-five-page booklet on Americanism published in 1936. One San Francisco columnist, who identified Agnes as "a patriotic American and a brilliant author," described her book succinctly: "In a series of simple questions and answers she puts the elementary principles of this republic before the reader in such shape that understanding is easy." And public lecturer Aline Barrett Greenwood reportedly told her California audiences that the *American Primer* "should be in the hand of every thinking American."

During the spring of 1934, Agnes and Lou Henry exchanged letters and notes on social and political issues, and on occasion they visited in each other's homes. One day, probably in late April, Lou left behind at the Cleaveland residence a package containing a copy of *De Re Metallica*, a book on mining engineering first published in 1556, which Lou and Herbert Hoover together had translated from Latin to English and republished in 1912. Agnes was deeply moved by this gesture of Lou's friendship. She told the Hoovers, "When I opened the 'bric-a-brac' after Lou left the other day, I sat down and wept. I have to confess it. That's all I can say by way of thanks." It inspired her, she said, "to give the Political Science Club the best American doctrine I've ever expounded."

That spring also found Herbert Hoover hard at work on a manuscript, "The Challenge to Liberty," in which he outlined his public philosophy. He sent drafts to close friends and writers for comments. Lou Henry also sought feedback from her trusted friends. "I am taking the liberty of sending you a copy," Lou wrote to Agnes on May 18, "because I am quite sure you will be interested in it and because I would very much like your opinion of it, as I am sure my husband would." She wanted Agnes to ascertain "what Mrs. Average Citizen would think of it and get from it. (Since you frequently affirm that you are Mrs. Average Citizen, and at the present moment I don't want to contradict you!)." Of greater importance, Lou wanted to know whether Cleaveland thought it was "a good book to give the American people. And then do you think it is a good book to give the American public *now*?" She told Agnes, "I would just like your unbiased opinion. For that reason I am not telling you what I think about it. But I am enclosing in another sealed envelope my opinion."

Presumably, Agnes read the manuscript with great pleasure, adding this task to the myriad other opportunities she grasped to speak or write about her political convictions. On

May 16, she began a weekly column in the *Carmel Villager* (later *The Town Crier*), in which she raised questions about the New Deal and the direction it was taking the nation. Later that month, she debated "a rather well known Berkeley citizen who has socialistic leanings" on the question "Is wealth in America too largely concentrated in the hands of a few?" In mid-June she addressed the Berkeley Republican Club, the Methodist Women's Conference in Stockton, and delivered her Wealth in America talk on a local radio station. She then headed south to Los Angeles to fill several other speaking engagements.

Agnes had barely reached her destination when she learned of Eugene Manlove Rhodes's death on June 27, 1934, at his home in Pacific Beach north of San Diego. In *No Life for Lady*, she told how she received this news: "I was stepping into my car to drive to Pacific Beach to visit him and 'Mary and Martha,' his inspired name for his wife, when the telegram announcing his death was handed to me." Remembering the good times they had shared, she added, "That he called me friend goes far toward balancing life's ledger with its violently fluctuating record of triumphs and disappointments." She would later pay tribute to Rhodes's genius at a memorial service held in New Mexico and in the pages of her autobiography.

By July 7, Agnes had delivered five talks and spoken to key women (some at Lou Hoover's behest) on issues of education and good citizenship. Then, following a month-long visit to Datil, Agnes spent a good deal of time giving public lectures, defending the Hoover name, and working to defeat Upton Sinclair in his bid to become governor of California. "Speeches are coming up," she wrote to Lou Henry on August 20, "and I am trying to think up new ways of saying the same thing." She was "thrilled to the core" to learn that Herbert Hoover's book was about to appear in bookstores.

In *The Challenge to Liberty*, published in mid-September, Hoover elaborated on his ideological views and offered to the public a nonrancorous and "theoretically abstract criticism of the New Deal." He warned against collectivism and "national regimentation." And he stressed that men everywhere "possessed fundamental liberties apart from the state, that they were not the pawns but the masters of the state."

Although Hoover's book soon made the best-seller lists, it was not "welcomed universally." Among its detractors were J. Stitt Wilson, Berkeley's former socialist mayor, and Dr. Eldred Vanderlaan, founder of the Humanist Society of Berkeley. Like other critics of the former president, Wilson and Vanderlaan believed that he had done nothing to relieve suffering during the Depression. Agnes had listened to both men rail against Hoover in public meetings. Her written rebuttal to Wilson's speech, which she called "inflammatory," was printed in the *Berkeley Gazette*. Likewise outraged by Vanderlaan's attack—"the most vindictive thing I have ever listened to"—she asked to reply. Vanderlaan obliged, and the following Sunday Agnes spoke before his congregation in defense of Hoover's record and *The Challenge to Liberty*.

During the fall of 1934, Agnes became heavily involved in one of California's most sensational gubernatorial campaigns—one that pitted the socialist writer Upton Sinclair against the Republican Frank Merriam. By this date, Cleaveland had shed her own infatuation with socialism. Her turn to conservative politics probably dates from the 1920s when she witnessed homesteaders pouring into west-central New Mexico to take advantage of the Stock Raising Homestead Act of 1916, which allowed 640-acre homesteads. In a land of little rain, these newcomers had a tough time raising enough crops to survive; many became discouraged and left. Others who stayed earned the animosity of local ranchers after they cut the stockmen's fences, butchered their steers, and hauled water away from their wells. Ranchers also had to pay taxes for the schools and roads that mainly benefitted the homesteaders. Initially sympathetic to their plight, Agnes later held homesteaders responsible for many of her brother's financial problems.

Like other politically conservative women of her era, Agnes embraced nonpartisan action to redress social and political evils. Yet in state and national elections, she supported the

Republican Party, believing that it best represented traditional American values. Republicans, in fact, had dominated California politics since the 1890s, although a resurgent Democratic Party would seriously challenge the Republican organization during the Depression.

In the 1934 gubernatorial election, Earl Warren, Alameda County district attorney and new state chairman for the Republican Party, led the fight against Upton Sinclair. Warren believed that Sinclair's much ballyhooed "End Poverty in California" program was responsible for the droves of unemployed and penniless migrants flocking to the state. And he warned voters that "Sinclair's election would represent the end of civilized democracy in California." He also framed the election as a battle, not between Republicans and Democrats, but between capitalism and socialism, Americanism and communism.

Agnes worked hard to assure Sinclair's defeat. She gave speeches on Merriam's behalf; wrote letters critical of Sinclair; and served as Alameda County chairman of speakers for meetings and radio broadcasts. In a letter written to her daughter Mary (while staying overnight at the Hoovers), Agnes described what her duties as chairman entailed. "All speeches have to go over my desk [at Republican headquarters in Oakland] for censorship and all speakers are invited by me." (She also wrote speeches for people who could not write their own.) In this campaign, she added, "there are no party lines. It is Carl [*sic*] Marx versus the Constitution of the United States. So we have many Democrats as well as Republicans on our side of the fence." Voicing a sentiment held by many other voters, Agnes wrote, "Our work is not to elect somebody but to defeat one man." She ended the letter on a happy note: "It is time to go upstairs and meet Herbert. He gave me an autographed copy of Challenge to Liberty. Am I puffed up?"

Following Sinclair's defeat at the polls in November, Agnes resumed her role as public lecturer, giving talks to women's clubs and on the radio. She also attended a variety of civic events. One of the most memorable was the banquet held in mid-January 1935 to honor Amelia Earhart, who days earlier had completed a solo flight from Honolulu to Oakland. Numerous Washington dignitaries, including President Roosevelt, sent congratulatory messages. Former President Hoover was present in person to praise her accomplishments. But what made the evening special to Agnes was the introductory speech delivered by Lou Henry. "I had never heard you before," she wrote to her friend the following day, "and I was pleased as punch." In Agnes's estimation, "Your introduction of Amelia Earhart was one of the most graceful and at the same time most meaty little talks I've heard in a long time."

Agnes opened the fall 1935 club season as newly elected president of the Alameda County Federation of Women's Clubs, an affiliate of the California Federation of Women's Clubs. The CFWC, founded in 1900, was "an umbrella organization" that united hundreds of clubs throughout the state. CFWC conventions gave clubwomen the opportunity to exchange ideas and plot strategies for united action. Local clubs sponsored a wide array of projects, such as beautification of communities, protection of the giant redwoods, immigrant education, support of public libraries and children's playgrounds, and lobbying of the state legislature for political and social reforms.

Cleaveland's agenda as president of the Alameda County Federation (a unit that represented more than four thousand women) reflected her own passion for the twin causes of good government and good citizenship. At a June meeting in her home, she outlined a program of Americanism that would be launched in the coming months—with emphasis placed "on privileges of franchise" and "other American practices of government which make for good rule." And in a radio address delivered in September, she announced that the Alameda County Federation was beginning its club year with a session devoted to the American Constitution.

Before the year ended, however, Agnes had become the center of attention of a new patriotic women's organization, Pro America, National Organization of Republican Women. The history of Pro America has yet to be written; unexamined until now have been Agnes's papers relating to its emergence as a national organization. These documents make clear, however, that her association with Pro America led to much personal distress on her part as well as dissension within the leadership of the organization.

Pro America was founded in Seattle in 1933, its members "dedicated to clean government and to preserving the American form of government." It had enlisted the aid of Edith Kermit Roosevelt, widow of the late President Theodore Roosevelt, who became Pro America's honorary president. Elizabeth Hanley, Pro America's national organizer, also tried to secure Lou Henry Hoover's support. Writing in December 1932, she asked the first lady's help in forming a chapter in California after the Hoovers left the White House. Although Lou Henry declined to become actively involved in Pro America at this time, both she and her husband kept informed of its affairs and in due time introduced Hanley to Agnes Morley Cleaveland.

Agnes was thrilled when asked to take a leadership role in this new organization. In a letter to her daughter Loraine, written in November 1935, she exclaimed: "Well, I seem to have gone off the deep end! Something is going to pop. I am going to accept the chairmanship of the Western Dist. of Pro-America." She described Pro America as an association that "hopes to coordinate all other patriotic women's organizations—just to be a sort of clearing house for women's united effort to save America." With chapters in Washington, Oregon, Idaho, and Montana, Agnes believed her main duty would be to travel the countryside giving pep talks. "Right up my alley," she enthused. Still, she had only a hazy idea about Pro America's goals and principles. But, because Herbert Hoover had "proposed my name [for the chairmanship] . . . I couldn't refuse. He told the Executive Committee who consulted with him that I was the one woman of his acquaintance who was 'qualified for the job from every angle.'" Later she would admit that she accepted her new role in Pro America too hastily.

Events unfolded rapidly. Hanley traveled to California early in December to meet with the Hoovers and then with

Cleaveland. She also spoke to a nonpartisan organization of women in Oakland, known as the Constitutional Government Group, to enlist their support for the organization. According to Agnes, Hanley presented Pro America as "designed primarily to interest women who had never taken an active part in public affairs," and she said that while Pro America was "based upon Republican principles" they need not be "shouted from the housetops." Indeed, Pro America's slogan was "Principle before Party." Hanley also announced at this meeting that Cleaveland had been "drafted" as national president.

In January 1936, Agnes traveled to Seattle, where she spoke at a large meeting of Pro America and formally accepted the presidency. After consulting with other leaders, she left Seattle with the firm belief that Pro America was nonpartisan, that it encouraged women of both parties to join its crusade to unseat President Roosevelt and other New Deal bureaucrats.

Upon returning to California, Agnes went immediately to Los Angeles, where she met with Edith Van de Water, Republican National Committeewoman from California, and with a group of women prominent in club and civic affairs, to explain the purpose of Pro America. According to one of Lou Henry Hoover's friends, Agnes made "a very good impression." She described Cleaveland's presentation as "a stunning piece of work. It was really brilliant."

On January 23, Agnes resigned her position as president of the Alameda County Federation of Women's Clubs to devote her entire time to the activities of Pro America. In her resignation statement she referred to her past efforts "to awaken our people to the danger confronting them from the breakdown of the spirit of initiative, independence and self reliance." Now the time had come to align herself "with an organization which was pursuing all of these objectives in a bigger way and thus, perhaps, extend my influence in this hour of my country's crisis."

But Agnes already was beginning to question the use of the subtitle, National Organization of Republican Women, in Pro America's name. She could not truthfully say that the movement was either Republican or nonpartisan. For a time, she accepted the word Republican "as synonymous with anti–New Dealism, rather than synonymous with Party Machinery." In trying to clarify her thinking, she held interviews with Earl Warren and other leaders in the Republican Party. She was told that as a nonpartisan organization fighting the New Deal, Pro America "could be valuable politically." But as an independent Republican group, "not subject to party control," it might become "a political nuisance." Agnes came to believe that Pro America officials in the North meant to proceed as an independent unit without necessarily adhering to party policies, which, she was convinced, would undercut the strength of the Republican Party.

Cleaveland's subsequent attempts to jettison "Republican" from Pro America's name and turn the association into a strictly nonpartisan group that stressed patriotic education led to stiff resistance from Elizabeth Hanley and other Pro America leaders. By early March, they were calling for Agnes's resignation as national president. Cleaveland came out of this in-fighting battered in spirit, or, as she wrote to Loraine after she had lost the battle and relinquished the presidency, "I've been through a lot of mental stress and strain but am emerging."

Mercifully, Lou Henry Hoover provided Agnes with much needed emotional support. Agnes had talked at length with Lou Hoover in early March about her problems with Pro America. When Hanley wrote to Hoover on the tenth complaining of Cleaveland's "controversial attitude" and expressing the fear that "poor Mrs. Cleveland's [*sic*] blundering may destroy a very fine idea," Hoover came to Agnes's defense. She told Hanley that after she (Hanley) had presented the idea of Pro America to the Hoovers, Lou Henry had remarked to Agnes and to others that "her effort here and yours in the North were almost parallel. . . . From the first conversations she had with you and the earlier descriptions from me she thought that you were on quite the same track that she herself had been following this couple of years." Diplomatically, Hoover expressed her belief that both methods, partisan and nonpartisan, "are almost necessary in reaching the different types of women who are becoming more and more interested in these problems."

Agnes's relinquishment of the presidency in mid-March did not end her troubles with the organization. Some weeks later, she received word that Sarah Battle, a Pro America activist and lecturer, had said in a public meeting in San Francisco that Cleaveland had accepted a $10,000 bribe to change Pro America into a nonpartisan organization. Several notable people, including Ben Allen, an intimate of Herbert Hoover, advised Agnes to sue Battle for libel. "My family insisted upon it," Agnes avowed. When faced with legal action, Battle issued a formal retraction of her statement.

Knowing that both Hoovers looked kindly upon Pro America, Agnes was doubly grateful for Lou Henry's steadfast support. "It was dear of you," Agnes wrote on April 18, 1936, "to let me come down the other day when my spirits were at such a low ebb. I felt as though I had foozled things horribly—but I've decided against crying over spilt milk and am now looking about for more trouble to get into."

Some high-profile California Republican women would continue to support Pro America after Agnes left the organization. Among them were U.S. Congresswoman Florence Prag Kahn and Ruth Comfort Mitchell, author, playwright, and long-time Republican campaigner. Elected president of the California chapter of Pro America in 1939, Mitchell would make a minor splash in the literary world when her anti–Grapes of Wrath novel, *Of Human Kindness*, was published in 1940.

Meanwhile, instead of trouble, Agnes Morley Cleaveland found outlets for her pent-up energy as a grassroots political activist. She became "a ward doorbell ringer" during the fall 1936 presidential campaign, which pitted the Republican Alfred M. Landon against the incumbent Franklin D. Roosevelt. She also spoke at anti–New Deal rallies and debated prominent Democrats on the merits of each party's platform.

Landon's defeat in November did not in the least dampen Cleaveland's political fires. When Roosevelt sent to Congress, in February 1937, his Judiciary Reorganization Bill (better known as the court-packing bill), she joined in the nation-wide protest. Alarmed by this threat to the integrity of the Supreme Court, a small group of Bay Area Republicans, which included Agnes, Ben Allen, Mrs. Lee Breckenridge Thomas (head of the women's division, Alameda County Republican Committee), and former U.S. Congressman Ralph R. Eltse and his wife, the writer Oma Almona Davies, quickly organized a mass meeting in Berkeley to denounce the president's proposal. More than a thousand people attended, according to Agnes's estimate, "many being turned away." Afterward, people clamored for further action. In response, several members of the original Bay Area group (Agnes among them) met and organized the League for Supreme Court Independence. This association continued to stage public protests until, in the face of overwhelming opposition, the president conceded defeat and in August dropped his attempts to enlarge the Supreme Court.

Agnes Morley Cleaveland's political activism carried over into 1938 when she became the principal organizer of the Northern Division of the California Council of Republican Women. This organization "had its inception," Agnes recalled, with a visit of John M. Hamilton, chairman of the Republican National Committee, to San Francisco in January. At this time he spoke to a small group of women activists about the objectives of the National Federation of Women's Republican Clubs, then being put together by Marion Martin, assistant chairman of the Republican National Committee. In her travels across the country, Martin had encountered scores of women's Republican clubs that often acted independently of the Republican Party machinery. She wanted to strengthen party loyalty among women "both to help the party and to advance women in politics." As she envisioned it, the federation would serve as a national umbrella organization that would have "jurisdiction over state federations of individual clubs."

In February, Agnes embarked on a grueling schedule to organize the northern counties of California. Within a span of three months, she traveled more than five thousand miles by auto and made one trip by air. She also helped arrange Marion

Martin's visit in April, when Martin helped launch the California Council of Republican Women, soon to become an affiliate of the National Federation of Women's Republican Clubs. At a dinner held in Oakland on the twentieth to honor Martin, Agnes sat at the speakers' table, along with Lou Henry Hoover, who presided, Marion Martin, Edith Van de Water, and Mrs. Charles H. Rowe, chairman of the affair.

Agnes's work as organizer of the Northern Division signaled the end of her public advocacy of nonpartisan politics in favor of party loyalty. She now agreed with Marion Martin that "women had a civic responsibility to work within the parties." Agnes believed that nonpartisan and bipartisan action had been necessary in the early days of fighting the New Deal. But, under the present circumstances, the two-party system had to be strengthened. She feared that multiple parties and a proliferation of special interest groups (as found in California) would lead to political chaos and then dictatorship. "Only through a fairly even division of the people into two major groups," she avowed, "can steady, systematic and orderly progress be achieved."

On June 29, the day Cleaveland's term as organizer ended, she reported that Republican organizations had been set up in twenty-two of the thirty-six counties in the Northern Division. With her job successfully completed, she told Lou Henry that she hoped "to revert to private life and conduct a one-man sniping campaign at the New Deal." Nonetheless, Cleaveland stayed active in the California Council and later would serve on the board of directors of the Northern Division.

After a brief trip to New Mexico, in August 1938, Agnes continued her political activism. She campaigned aggressively throughout the fall for Philip Bancroft, the California Republican candidate for the U.S. Senate. Bancroft, a well-to-do Walnut Creek rancher (and son of historian Hubert Howe Bancroft), was a far-right conservative and rabid anti–New Dealer. His opponent, Sheridan Downey, an Atherton attorney and former running mate of Upton Sinclair, was a far-left Democrat, who championed the "Ham and Eggs" movement—an effort to provide $30 every Thursday to unemployed Californians over the age of fifty. After Downey trounced Bancroft at the polls, Agnes wrote to her daughter Mary: "I worked to the limit of my strength, time and emotion."And she added, "I slept 24 hours out of 36 the days following the election and am beginning to feel normal for the first time in months."

Cleaveland's exhaustive agenda as civic and political activist leaves the reader to wonder when she had time for family or other activities. Almost every summer she spent up to four weeks in Datil. Her daughter Loraine, who lived in Hollywood with her banker-husband George Keffer, sometimes accompanied her. Together, they entertained as summer guests such noted writers as Laura V. Hamner (author of the classic *Short Grass and Longhorns*, 1943) and Hilda Faunce Wetherill (author of *Desert Wife*, 1928), as well as Agnes's cousin, the archaeologist and Maya scholar Sylvanus Griswold Morley, and his wife, Frances.

But family tensions sometimes spoiled any image of Datil as an idyllic haven from worldly cares. In August 1934, for example, Agnes confessed to Lou Henry that she had "suffered great mental stress and strain" while in New Mexico. The recent marriage of her youngest daughter, Mary, to Claus Wohlers of Gold Creek, Montana, undoubtedly caused much of her distress. Mary was as headstrong as her mother and had married Claus against her father's wishes and without informing her mother. Agnes learned of the event from a newspaper clipping. Newton, then overseeing dredge-mining operations in the Gold Creek area, had been invited to the wedding but refused to attend.

By this time, however, Agnes and Newton's own marriage seems to have been deeply troubled. He often was away on work assignments, leaving her to manage the household and to create a public life for herself separate from his own. Their physical separation evidently contributed to their growing emotional detachment. In October 1934, Agnes wrote to Mary (after mother and daughter had reconciled): "Now about Pop.

He isn't worrying me any more. No use to let him worry you. I've pulled through 35 years and since the first 100 years are the hardest, I'm past the third-of-the-way mark." Later that year, Agnes told Mary that the house in Berkeley had a For Sale sign on it. "I have come to the point of wanting to sell. I don't like it here alone so much of the time. . . . one of these days I'm walking out of it and not coming back!"

The Cleavelands' marital troubles stemmed from more than their physical separation. But gaps in the family history force one to speculate about causes—was it Newton's male assertiveness or Agnes's stubbornness that was more to blame? Or was it, as suggested by their son, that "Pop is certain that he is a master-mind"—but, then, he said the same thing about his mother. And so, we are left with two forceful personalities who seemed increasingly to be at odds.

Still, the Cleavelands followed typical family rituals. Agnes traveled to Gold Creek the summer of 1935 to be present for the birth of her first grandchild, Beth Wohlers, on June 23. She again visited Montana in December to spend Christmas with Newton and the Wohlers; she carried with her on this trip a package from Lou Henry Hoover, which evidently contained the baby blanket that the former first lady had knitted for Mary's baby.

Amazingly, amidst family concerns, travel, club work, and political chores, Agnes found time for her writing. Granted, many of her compositions in the 1930s dealt with civic or political issues; but she also mulled over ideas for a grander, literary effort. By the time of the Bancroft campaign, she was well along in writing her memoirs. Although she often had been the focus of attention on the California political scene, nothing prepared her for the national acclaim she received after publication of *No Life for a Lady*. And she loved every minute of being in the limelight.

Ada McPherson Morley. Courtesy New Mexico State University Library, Archives and Special Collections, MS00250004.

William Raymond Morley. Courtesy New Mexico State University Library, Archives and Special Collections, MS00250001.

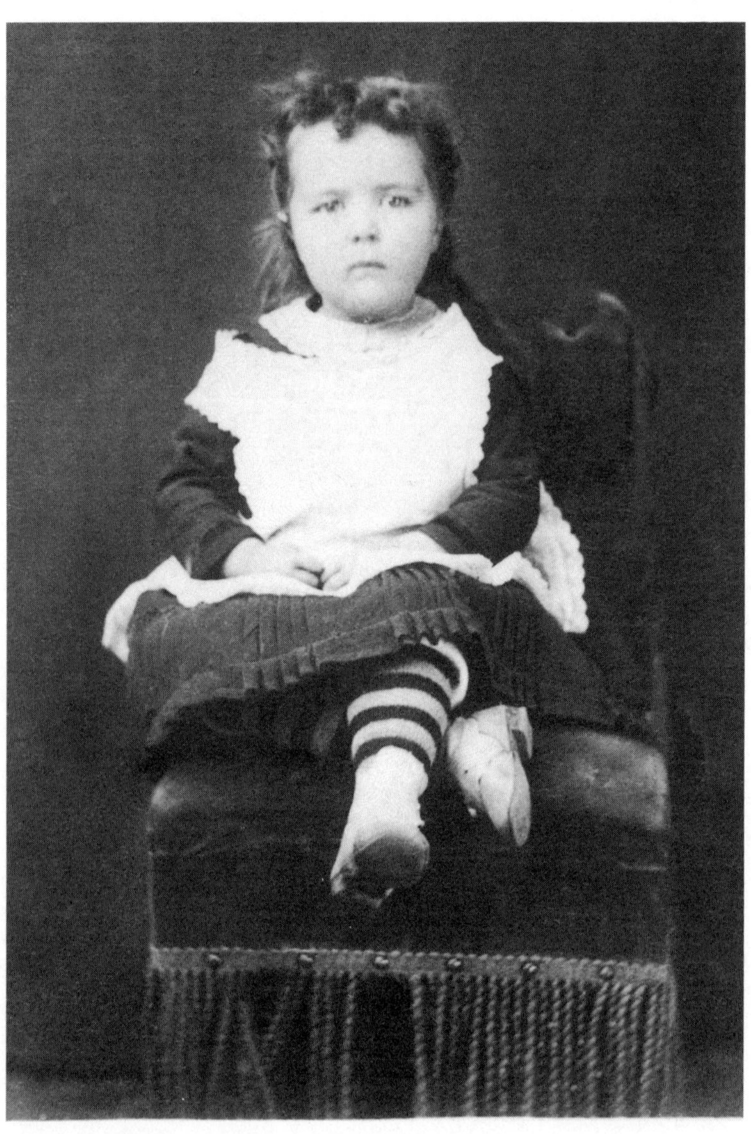

Agnes Morley, three-and-one-half years old. Courtesy New Mexico State University Library, Archives and Special Collections, MS00250006

Levi Baldwin Place, original Datil post office. Courtesy New Mexico State University Library, Archives and Special Collections, MS00250041.

Lora and Agnes Morley. Courtesy New Mexico State University Library, Archives and Special Collections, MS00250008.

Agnes Morley, "A Jekyll-Hyde Life," photographs taken while attending college in Ann Arbor, Michigan. Courtesy New Mexico State University Library, Archives and Special Collections, MS00250009.

Agnes Morley on a grizzly bear hunt, 1895. Courtesy New Mexico State University Library, Archives and Special Collections, MS00250048.

Ray Morley at Columbia, 1901. Courtesy New Mexico State University Library, Archives and Special Collections, MS00250018.

Agnes Morley Cleaveland and son Norman on the Swinging W Ranch, 1903. Courtesy New Mexico State University Library, Archives and Special Collections, MS00250073.

Fred Winn painting Agnes Morley Cleaveland's portrait, Norman in foreground. Courtesy New Mexico State University Library, Archives and Special Collections, MS00250023.

Agnes Morley Cleaveland on the Swinging W Ranch. Courtesy New Mexico State University Library, Archives and Special Collections, MS00250057.

Ray Morley's Kindergarten Outfit, taken by Dane Coolidge in 1918. Courtesy New Mexico State University Library, Archives and Special Collections, MS00250132.

Navajo Lodge, Datil, New Mexico, 1940. Courtesy Library of Congress, LC-USF34-035886-D.

Interior of Navajo Lodge, Datil, New Mexico, 1940. Courtesy Library of Congress, LC-USF34-035893-D.

Agnes Morley Cleaveland, portrait accompanying publicity for *No Life for a Lady*. Courtesy New Mexico State University Library, Archives and Special Collections.

Agnes Morley Cleaveland's ranch at Jack Howard Flat. Courtesy New Mexico State University Library, Archives and Special Collections, MS00250110.

Garden at Jack Howard Flat. Courtesy New Mexico State University Library, Archives and Special Collections, MS00250113.

Agnes Morley Cleaveland contemplating rabbit brush on Jack Howard Flat. Courtesy New Mexico State University Library, Archives and Special Collections, MS00250015.

Agnes Morley Cleaveland at Jack Howard Flat, circa 1950. Courtesy New Mexico State University Library, Archives and Special Collections.

CHAPTER 5

No Life for a Lady

AGNES Morley Cleaveland liked to tell stories—whether on paper or in front of an audience. Probably it was only a matter of time before she put all of her stories into a coherent narrative and sought a publisher. But two literary friends played important roles in bringing her masterpiece to fruition. The first, Eugene Manlove Rhodes, had chided her for abandoning her writing career after their collaboration on "The Prodigal Calf." Evidently, he kept after her to write the stories she knew the best, about the range cattle industry of west-central New Mexico. Shortly before he died, Rhodes said to a mutual friend: "You tell Agnes Morley Cleaveland I'm goin' to haunt her from the other world till she goes back to writing."

The second important literary figure was Conrad Richter, a Pennsylvania-born writer who had not yet published his first novel when he and Agnes met through an exchange of letters in 1934. Conrad and his wife, Harvena, had moved to Albuquerque, New Mexico, in 1928, in hopes that a drier climate would restore Harvena's health. Providentially, the Southwest provided Richter with material "for his first prominent achievements in fiction." Between April 1934 and January 1936, he published nine stories in the *Saturday Evening Post* and the *Ladies Home Journal*, each with a Southwest motif.

Agnes read Conrad's "Long Engagement" (later retitled "Long Drouth") in a June 1934 issue of the *Saturday Evening Post* and was struck by his skill in capturing "the tragedy and the majesty of her part of New Mexico." She wrote to him at some length, congratulating him on his success and apparently describing several colorful episodes in the Morley family history. In return, Conrad thanked her for her "fine, generous letter." He had known of Agnes for some time, he said. "Kind letters come in about my stories but it is a red-letter day when I am able to draw the approval of one who knows the field of which I am trying to write like you do."

Agnes and Conrad continued to exchange letters, she writing "delightful authenticities" of her life in the Datils, he describing his recent literary work, assuring her that his depiction of a young railroad engineer in one of his stories was not modeled on her father. In due time, Richter went to see Agnes during one of her returns to Datil. They talked for hours. Especially memorable for Conrad was Agnes's vivid account of the conflict that raged between the cattlemen and the homesteaders who flocked to New Mexico after passage of the 640-acre Stock Raising Homestead Act. This contention subsequently became the background for his well-received first novel, *The Sea of Grass* (1937). Conrad readily acknowledged his indebtedness to Agnes. After Edward Weeks, writer and editor of the *Atlantic Monthly*, learned that some "oldtimer" had suggested to Richter that he "give the cowman's side of the picture" and that "The Sea of Grass was the result," he asked Richter if this were true. Yes, Richter replied; and he identified "Agnes Morley" as the old-timer.

At the end of their first lengthy interview, Agnes had told Conrad, "This story [cattlemen vs. homesteaders] is yours. I do not feel equal to writing it myself and I am offering it to you as a gift." But Conrad felt an obligation to repay her. So, as Agnes later told a reporter, Richter announced that "he was going to hound me until I wrote my own book and stopped giving away what the writing craft would call 'my material.' I must say he did a thorough job of it in person, in letters from wherever he might be, Albuquerque, the West coast and in the East, and by relayed messages." With a bit of wit, she concluded, "He harped on that one theme until I wrote the book in desperation to get Conrad Richter out of my hair."

In mid-November 1938, Agnes received a letter from Alfred A. Knopf, Richter's publisher, approving the ten-thousand-word sample of her book that she recently had sent and asking that she send the rest immediately. None of her papers reveal when she had started work on the manuscript. But in a letter written to her daughter Mary the day after she heard from Knopf, Agnes reported that she had completed half of the chapters and expected to finish the rest by Christmas. She was much encouraged by Knopf's response. The letter seemed to assure "the book's acceptance if it all holds up to the sample. I had sent what I considered the poorest sample in order to be more than sure they'd take it, if at all, so I feel awfully sure it will click."

Mary and her two infants, three-and-a-half-year-old Beth and three-month-old Norman, arrived at the Cleaveland home, evidently in time for Christmas, with plans to remain two or three months "to escape the cold in Montana." The day after New Year's 1939, Agnes wrote to thank Lou Henry Hoover for the box of raisins the Hoovers had sent for the holidays. She also said that one day she would call on the Hoovers with her daughter and granddaughter since "Mary very much wants Beth to be able to say, in later life, that she met Herbert Hoover."

Agnes soon resumed her club activities, speaking at a conference of Women Defenders of America (a patriotic organization) in mid-January, to the Political Science Club in early February, and attending the annual convention of the California Republican Assembly on February 11, as a delegate from Alameda County. She must have waited impatiently to hear from Alfred A. Knopf. And when a letter finally arrived, she would have been thoroughly disappointed—the firm would not be publishing her manuscript.

In mid-May 1939, Agnes sent a sample of her story, now entitled "We Made Our Own Tracks," to H. S. Latham, vice president of the Macmillan Company. In a cover letter, she reminded him that he had looked over her "American Primer" manuscript four years earlier while he was in Berkeley. Declining the opportunity to publish it, he had kindly suggested that she try her "hand at something else." Thus, she told Latham, "I herewith enclose 'something else.'" Eight weeks later she received Latham's politely worded letter of rejection. (After *No Life for a*

Lady became a huge success, Agnes placed Latham's letter in a scrapbook filled with laudatory reviews, but she pasted the offending missive face down, making it difficult to read.)

Quite likely, Agnes set the manuscript aside for awhile and concentrated on civic and family responsibilities. She spent much of the summer of 1939 in Datil, where she made arrangements to build a log house for Morley on Jack Howard Flat. In mid-June, while driving at night on the road between Socorro and Magdalena, Agnes escaped without injury when her car was side-swiped by a truck, although her car was heavily damaged. In the fall, Agnes's days were taken up with meetings, speeches, and work on behalf of the California Federation of Women's Clubs. In her role as the Federation's chairman of American citizenship, she wrote the first of a series of essays that would appear in the monthly *California Federation News*.

Agnes continued to speak at women's meetings during the first half of 1940. Her topic usually focused on American citizenship, although occasionally she told of her childhood adventures in territorial New Mexico. She again spent the summer months in Datil, staying with Morley in the new house on Howard Flat. Loraine joined them for part of the summer, as did Bill Morley, the late Ray Morley's grandson.

Agnes returned to Berkeley during the fall presidential campaign, in which the Republican Wendell Willkie unsuccessfully attempted to oust Franklin D. Roosevelt from the White House. With all of Europe inflamed in war, however, Agnes found herself supporting Roosevelt's foreign policy. In a letter to Willkie written on September 6, she voiced her impatience with those who questioned whether the president was adhering to the "democratic process" in his foreign procedures. In Agnes's view, the Founding Fathers had given the chief executive wide-ranging powers to deal with totalitarian countries who posed a threat to the United States. "Hitler and Hitler alone will get us in or keep us out of war," she continued. "Our preparation must be a willingness to fight and to let him know it. We [can't] go bleating around about democratic processes when we are talking to the Great Bully of all time. Let's fight

Roosevelt where he is vulnerable and that's his domestic fiasco but not aid Hitler in fighting him when he is right."

Agnes also believed (along with other conservative Americans) that among the gravest dangers facing the United States was the Communist Party, which meant to replace the nation's capitalistic system with "Soviet-style socialism." The Communist Party, in fact, obtained its greatest influence in the United States during the decade of the thirties when many men and women were drawn to its ranks because of the tragedies wrought by the Great Depression. With the nation's economy in shambles, the "Soviet experiment seemed vital and alive."

Throughout the decade, some figures in California ran for political office on the Communist Party ticket. Still, this precedent did not stop leaders of the California Federation of Women's Clubs from adopting a resolution, introduced by Agnes Morley Cleaveland, in September 1940, demanding that the Communist Party be removed from the ballot "as not fulfilling the requirements of a bona fide political organization." The women asserted that the party was "a branch of a foreign agency," which meant to establish (by violence if necessary) "a worldwide order antagonistic to the principles and system on which this country was founded."

What makes Cleaveland's condemnation of communism of more than passing interest is her relationship with young Kenneth May, a family friend who made headlines on September 27, 1940, after his father, Professor Samuel C. May of the University of California, disowned him for being a member of the Communist Party. The Mays had lived in north Berkeley since 1921. As a youth, Kenneth met Mary Cleaveland on the courts of the Berkeley Tennis Club; although he was two years younger than she, they became close friends. Mary later recalled that Kenneth often visited in the Cleaveland home and that Agnes became quite fond of him—despite his liberal political leanings.

Kenneth joined the Communist Party while an undergraduate at the University of California, from which he graduated in 1936. After studying abroad for a time, he returned to Berkeley in 1939 to pursue graduate studies in mathematics while employed on campus as a teaching assistant. But changing world events the following summer would have a profound impact on his career.

After Hitler's rapid march across Europe and the fall of France, in June 1940, Americans worried about their own security. To prepare for possible entry into war, Congress passed a conscription law in September, the first peacetime draft in the nation's history. That same month California created a State Council of Defense and appointed Samuel May as its executive vice president (a controversial appointment, as it turned out). In this charged political atmosphere, Kenneth May went before a local school board, seeking a permit to use the Berkeley High School for a Communist Party function. The break between father and son came after the *San Francisco Chronicle* reported on the school board's meeting the next morning and identified Kenneth as a member of the Communist Party.

Critics of Samuel May have pictured his public condemnation of his son as a self-serving attempt to protect his position on the Council of Defense. Agnes surely interpreted his actions in this way also. Upon learning of the senior May's pronouncement, she wrote to him, suggesting that "Kenney" had learned his liberalism at his father's knee. She told May, as she repeatedly told others, that he had no right in cutting off his son. "You have known, as I have, that he was at least a 'fellow traveler' for several years." And, she said, "in the public mind you also were one. Your name is always mentioned in any discussion of the radical professors in the University. That you have taken no steps to counteract that impression until the hue and cry against your appointment to the State Defense Council would seem to justify the unhappy impression that you have thrown Kenney to the wolves."

Agnes also wrote to Kenney, sending him a copy of the letter she had written his father. She assured him, "I have a very real affection for you." But, she added, "naturally I think you are several varieties of an idiot in your political views as I have told you to your face on more than one occasion." On October 11, the Board of Regents of the university dismissed Kenneth

May from his teaching position because of his political beliefs. He later dropped his ties with the Communist Party, served in World War II with distinction, and went on to have a distinguished career as a mathematician at Carleton College in Minnesota and at the University of Toronto. Father and son apparently reconciled during the war years.

Meanwhile, Agnes continued seeking a publisher for her memoirs. In the late spring of 1940, she had met Eugene Cunningham, a writer of western fiction, at a literary luncheon in Berkeley. After telling him of her manuscript, he urged her to send it to Houghton Mifflin because "they liked westerns." And he was right. To capitalize on the nation's growing infatuation with western culture, the firm had published the works of several noteworthy western women authors, including Mary Austin's *The Land of Little Rain* (1903), Elinore Pruitt Stewart's *Letters of a Woman Homesteader* (1914), Anne Ellis's *The Life of an Ordinary Woman* (1929), and Mary Kidder Rak's *A Cowman's Wife* (1934).

Following Cunningham's advice, Agnes mailed a lengthy draft of her memoirs to Boston—and received no reply until late August. An editor made a half-hearted apology for the delay; her manuscript had been read two months ago, he explained, but "was shunted to a side track" when other business intervened. Still, the news was favorable. "Both our readers were interested in your story and felt that in places the incidents and the vitality of the telling were exceptional." He invited Agnes to submit the complete text.

Agnes went to work and quickly turned out a 60,000-word draft. More Houghton Mifflin people read it and liked it; they compiled a list of places where they thought it could be expanded or changed to benefit readers unfamiliar with the Datil territory. Moreover, Paul Brooks, the managing editor, took time while on a trip to the West Coast to discuss the book with the author. A reporter later described the editor's visit: "Brooke [*sic*] rang the bell of the Berkeley home when Mrs. Cleaveland was preparing to leave to address a woman's club in San Francisco. She telephoned to a friend to substitute for her. Then for two hours she listened to the editor tell her what was wrong with her 60,000 words." Most significantly, he urged her to make the book more autobiographical.

Agnes rushed off to Datil to handle the revisions—to be free of telephone interruptions, she said. Paul Brooks wrote to her there on November 27, 1940, offering words of encouragement. "The more I think of your book, the more I like it; and my enthusiasm is shared by the other editors here who have seen the sample manuscript." In two months of steady work, Agnes turned those 60,000 words into 150,000.

When it seemed assured that Houghton Mifflin would publish her manuscript, Agnes conveyed the good news to Conrad Richter. He gently reminded her of what he had said about her book "if you would work it over long and hard enough and keep sending it out. But you wouldn't do it for me—[would] only give me Bite-'em-Morley looks and letters. I'm delighted to know that you did it for Houghton Mifflin."

By late January of 1941, the editorial process was in full swing. Paul Brooks sent Agnes a list of questions posed by Kay Thompson, the editor assigned to work with her on the manuscript. "She is wonderful at this sort of thing," he avowed, "and has an unconcealed passion for your manuscript." And he added, "I am trusting you to once more detach yourself from your public by some subterfuge and let the world save itself for another few weeks."

When Brooks said further revisions would be made in the firm's offices, Agnes rushed to Boston to consult. "Imagine Boston editors trying to revise a manuscript on New Mexico life," she would later chuckle. Agnes worked well with Kay and her editor-husband Lovell Thompson. The best account we have of what took place during this editorial stage is from an interview Agnes gave to Socorro newsman Thomas Ewing Dabney for his article "Background for a Book," which appeared in the *New Mexico Magazine* in 1946.

> Chapter by chapter, Mrs. Cleaveland spread the manuscript on the floor of the Thompson home in Ipswich [Massachusetts]. All three, sitting on the floor, then proceeded to take the story

> apart, like a piece of machinery, and put it together again, with repairs and improvements. They chopped out about 30,000 words. Sometimes when Mr. and Mrs. Thompson made a suggestion, Mrs. Cleaveland said, "O. K. Rewrite that passage yourself." Replied they, "Not one word will we write. It's your story." . . . So sitting on the floor, Mrs. Cleaveland pecked out the rewritings on a portable typewriter.

None of the early versions of Agnes's book have survived. But four of Kay Thompson's letters, undated and filled with questions and comments, are found in Cleaveland's papers and provide additional insight into the editorial process. All seem to have been written after Agnes left Boston. In one message, Thompson complimented the author: "Have I said too often how wonderful it is to be able to indicate the kind of thing I'd like to have, and always always have you do it not only just as I wanted but with that extra something way beyond the specifications?" She also bluntly remarked: "I wouldn't have you change a word of that damned chapter. I only changed a couple myself. . . . I'm delighted with it."

A gifted editor, Thompson paid careful attention to the spirit of Cleaveland's stories as she rearranged sections of the manuscript. She told Agnes, for example: "All continuity in the last chapters is based on Ray; so I am busybodily splitting up everything you say about him and piecing it in to keep the movement as orderly as possible. I enclose accordingly the first page of chapter 38 for a very little re-doing; the rest of the chapter will follow along happily, without revision."

Elsewhere, she noted that five chapters, including the one entitled "Ray Lied Entrancingly," remained unchanged. She sent back the chapter on Gene Rhodes "because Paul would like more." And she suggested a new chapter, enclosing two pages describing how it might be handled. "This is just blocking out, you understand. I would like it to run to four or five pages of my favorite brand of your eloquence."

During the editorial process, the firm came up with the title for Agnes's book, *No Life for a Lady*. "We Made Our Own

Tracks," the author's original title, became a subheading on the contents page. Paul Brooks made fleeting reference to the name change in his correspondence with Agnes: "As for the title, we contribute that about two thirds of the time. We've got to do something to justify our existence." The new title proved popular with reviewers but later caused some scholars to conclude erroneously that it reflected Agnes's ambivalence about her life as a rancher—that she struggled to reconcile society's expectations to live the life of a lady with the exigencies of ranch life. On the contrary, Agnes gloried in her life as a rancher; she loved riding horses, working cattle, and experienced little difficulty in moving back and forth between her two lives.

In mid-April, Lovell Thompson contacted western artist Edward Borein to do the sketches for the book. And on May 12, Paul Brooks finally sent Agnes a contract to sign, a document that seems to have disappeared from the Houghton Mifflin files. He mailed the first copy of *No Life for a Lady* to the author on July 22 and expressed his delight with its appearance: "Since the format is all Lovell's, I can say with complete modesty that I think it's a knock-out." The book's jacket focuses on a pair of masculine boots that suggest a ranch hand sitting on the top rail of a fence; hills can be seen in the distance. Adding to the book's attractiveness are Borein's sixty-five sketches scattered throughout the text.

Meanwhile, on May 19, 1941, Agnes joined about two hundred old-timers and other guests in southern New Mexico to dedicate a monument at the grave of Eugene Manlove Rhodes, situated in Rhodes Pass in the San Andres Mountains near the site of Gene's beloved ranch. Among those who paid tribute to Rhodes's memory were New Mexico governor John E. Miles, Hugh Milton (president of the state college in Las Cruces where Gene had once been a student), Alan Rhodes (Gene's son), historian Walter Prescott Webb, and Agnes Morley Cleaveland. May Rhodes also was on hand to listen to the "anecdote swapping" that took place among old-timers who had ridden with Gene in his cowboy days.

Agnes spent the next two and one-half months in New Mexico, staying most of the time at the Cleaveland house on Howard Flat. On the evening of June 9, she appeared before an appreciative audience at the University of New Mexico in Albuquerque to talk about the writing of her book. By this date, Houghton Mifflin had awarded *No Life for a Lady* the firm's Life-in-America prize of $2,500, an honor noted by the reporter who covered Agnes's lecture. In an article published in the *Albuquerque Tribune* the day following her speech, he wrote: "When a 67-year-old woman writes her first and prize-winning book, that's news. And when the author is a once successful short story writer who stopped writing for 30-some years, that makes the story even more interesting." He recounted some of the stories Agnes told, describing her as "a grey-haired woman with a dry sense of humor." Toward the end of her talk, she told of her trip to Boston and how the publishers wished to test her authenticity. "[They] offered her a mouth organ after commenting upon her story of substituting for the fiddler with a mouth organ during a cowboy dance." "To the delight of last night's audience," the reporter continued, "Mrs. Cleaveland whipped out a mouth organ and showed how she settled the question of authenticity."

In mid-July, to mark the day on which his mother's book went to press, Norman Cleaveland flew his plane over the village of Datil and circled up the canyon. Local residents reportedly "shared in the thrill" that Agnes experienced watching his performance. Since there was no landing strip in the area, Agnes drove to Socorro, where Norman left his plane. It is presumed that he spent a few days with his mother at Howard Flat.

Bookstores across the nation received copies of *No Life for a Lady* in late July and in early August. Paul Brooks sent Agnes a telegram on August 5, about the day she returned to Berkeley: "Advance Sale Four Thousand[.] Readers Digest will run condensation in October issue paying one thousand dollars each to author and publisher[.] excellent review in New York Times today and review promised in Herald Tribune tomorrow."

In fact, the reviews were all an author could want. Fanny Butcher in the *Chicago Daily Tribune* (August 6) wrote: "Get out your notebook and jot down this admonition: 'Fanny Butcher says to run, not walk, to the nearest bookshop and hop onto a copy of 'No Life for a Lady.'" Rose Field in the *New York Times Book Review* (August 17) asserted: "This easy-flowing, informal volume written with humor and quiet wisdom, can be called both the autobiography of a person and the chronicle of a period and a section of the country." The *New York Herald Tribune Books* (August 10) devoted most of one page to Agnes's book, its headline shouting: "They Didn't Raise Cry-Babies in New Mexico, The Gallant and Authentic Saga of a Girl Who Grew Up on Horseback, When Men were He-Men." Stanley Walker wrote: "This personal history, telling of the ups and downs of a woman who was born in New Mexico in 1874 and is still going strong, will be immensely popular. . . . It is authentic, well written and, in many passages, downright charming. . . . Nothing quite like it has come out of the Southwest." Agnes's photograph and a picture of the book jacket accompanied Walker's review.

No Life for a Lady is made up of forty-five chapters or episodes: "some are long, some short; some dramatic, some hilarious." They comprise Cleaveland's memories of growing up on an isolated ranch in what is today Catron County and of the changes that came to the land. A born storyteller, she depicted her childhood in the Datils as one great adventure. Although she wrote to entertain, her knowledge of ranch life is evident on nearly every page. This combination of realism and humor assured the success of her book.

The early chapters record family history—Agnes's birth in Cimarron, her father's railroad building exploits, and his untimely death. Most of the book focuses on the rise and decline of the Morley fortunes after the family took up ranching in the Datils. Many of the stories deal with the exploits of ranch children (a brave and sturdy lot that learned self-reliance at an early age). With rollicking good humor, Agnes describes her rivalry with her brother Ray, their youthful high jinks, and childhood amusements. She also paints detailed pictures of ranch life—the twice yearly roundups, pulling bog in the rainy season, cutting hay, riding to Magdalena for supplies, and the myriad other tasks that spell hard work for the rancher. We learn that goods came from mail-order houses, illnesses were cured by home remedies, daily food was simple, and milk for infants came from range cows—at great risk to the milker.

Agnes recognizes that Ada Morley and her daughters led a freer and more exciting life than many of their women neighbors. Most ranchwomen lived in "terrible isolation and loneliness" while their menfolk were away looking after cattle. In contrast, the Morley women—beholden to no male head of household—moved freely on and off the ranch. Agnes also depicted the ranch as a classless unit that ignored differences between employer and employee. What counted was whether one made a good cowhand—whether one was "good enough to take along." It was a society without a double standard, where ranchwomen who did hard outdoor work like Agnes were treated as equals by the men they worked with.

Scattered throughout the book are wonderful descriptions of the New Mexico countryside. Here she writes of the glen where as a child she encountered a band of Navajos: "So many shades of green! The pine and spruce on the higher elevations, aspen a little lower down, piñons and juniper on the level stretches. So many brilliant hues of wildflowers on the valley floor! Lavender desert verbena, scarlet patches of Indian paintbrush, great blotches of yellow snakeweed. And above it a turquoise sky with white wooly thunderheads resting upon the mountain peaks." The description holds true more than a century after young Agnes watched the Indians ride away.

Cleaveland's stories were always about the ranch; she omitted almost everything of the life she led elsewhere. She says very little about her husband, Newton Cleaveland, and nothing directly about her children. Later, her son Norman would be identified as one of the two four-year-old boys who "flew at each other like a couple of cub wildcats" in the episode, "We Took It and Liked It." She also left out anything substantial about her stepfather, Floyd Jarrett, and totally ignored the part her mother's cousin Orrin McPherson played in operating the ranch after Jarrett left. Most noticeably (but understandably), she omitted painful, personal experiences, such as her teenage marriage to Mason Chase.

Although a superb storyteller, Agnes was not a meticulous scholar. Several important events are misdated, including the year the family settled in the Datils, the date Ray sold his ranches, and the year of Ray's death. And events that occurred in the two years preceding the Mason Chase episode are conflated into one, making it easier for the author to pass over that unfortunate affair.

Among the many strengths of this book are the author's vivid word images—almost picture perfect—of the life she lived as a rancher, the use of cowboy humor to keep the stories flowing, and the attention she pays to ranch children—girls and boys alike. Also commendable are Cleaveland's lack of pretension and ability to poke fun at herself, as well as her empathy for ranch families less fortunate than her own.

There are limitations, of course. One wishes that Cleaveland had been more forthright about her stepfather and her teenage marriage and that she had described the domestic side of ranch life (who cleaned, who cooked, who cared for clothing). Helpful also would have been a fuller description of the Morleys' interaction with the Alamo Navajo Indians and with the Pietown homesteaders.

Nonetheless, Agnes captured the essence of ranch life better than any other woman writer of her time. American folklorist and western literary critic J. Frank Dobie proclaimed that *No Life for a Lady* was "not only the best book about frontier life on the range ever written by a woman, but one of the best books concerning range lands and range people written by anybody." Today, more than sixty-five years after its publication, *No Life for a Lady* is still the best book we have about a western woman's ranch experience. No one has told the story better.

Dobie also had high praise for Mary Kidder Rak's *A Cowman's Wife*, published seven years earlier by Houghton Mifflin, and Nannie T. Alderson's *A Bride Goes West*, published by Farrar and Rinehart in 1942. These three authors—Cleaveland, Rak, and Alderson—shared a love for life on the range, but only Cleaveland had grown up on a ranch. Mary Kidder, born and raised in Iowa, graduated from Stanford a year after Agnes, and possibly the two established a friendship there (they corresponded in later years). Mary Kidder Rak took up ranch work at the age of forty when she and her husband, Charlie Rak, purchased a ranch in southern Arizona in 1919. Twenty-three-year-old Nannie Tiffany of West Virginia traveled west in 1883 with her new husband, Walt Alderson, to establish a ranch in Montana. Both Rak and Alderson tell engrossing stories of their lives on cattle ranches. Like Cleaveland, Rak became a capable cowhand and enjoyed working with cattle beside her husband. Alderson, on the other hand, spent more time at ranch headquarters, overseeing household tasks and later caring for her children. But she, too, on occasion helped out on roundups. All three were natural storytellers, but unlike Cleaveland and Rak, Alderson did not write the memoirs herself but had the author Helena Huntington Smith "take down her words" and turn them into a publishable manuscript.

Shortly after its publication, *No Life for a Lady* was reviewed in magazines and newspapers across the United States and in Canada. Journalists devised catchy titles to introduce the book: "Side-Saddle Tomboy," "Side-Saddle Cowpuncher," "They Had Guts!" "Poker, Bad Men and a Lady." Some California reviewers claimed personal knowledge of the author. In his column in the *Los Angeles Herald-Express*, the "Bookworm" wrote that he had "seen Mrs. Agnes Morley Cleaveland in action, bossing the conduct of a writers' club," and he would "testify that she rode herd on it most efficiently, with humor and force." Another reviewer, for the *Berkeley Gazette*, admitted: "We've known Mrs. Cleaveland for many years, we've heard her speak with dignity and with clear thought at dozens of luncheons and other meetings. But we never knew she could climb up on a corral fence, cup her hands over a harmonica, and play a tune that would make cowhands shout, 'Swing yer partners.'"

During the months of August and September 1941, the award-winning author was much in demand to speak before women's clubs and at literary events. At most club functions, including a dinner meeting of the California Writers' Club, she spoke about her book and how she came to write it. But at a luncheon held during the Writers Conference of the West (August 22–24) in Oakland, a conference attended by three hundred writers, she spoke on a more general topic, "Why Do Writers Write." Houghton Mifflin did its part to boost sales by inviting the public to meet Agnes and two of its other authors at a reception held in mid-September at the Sir Francis Drake Hotel in San Francisco.

On October 2, Agnes left on a whirlwind trip to the East, where she spoke at book fairs in Philadelphia and Boston, visited friends in New York City, and appeared as guest speaker at a tea in Washington, D.C., hosted by the League of Republican Women. En route, she autographed books in Albuquerque and in Chicago, where she had lunch with her cousin Walter Tibbles and breakfast with Arthur Kendall, the friend from Ann Arbor days who had let her ride a horse named Maud. At Fort Wayne, Indiana, the journalist "Slats" Logan boarded Agnes's train for a dinner interview and rode along with her as far as Lima, Ohio. In a book review entitled "Side-Saddle Cowhand," he described the author as "most gracious and deferential. She doesn't care about the trappings of a celebrity."

Agnes was one of several literary luminaries who spoke at the three-day Philadelphia Book Fair (October 6–8), an event that drew a crowd of more than two thousand to its displays and sessions. She also visited the Friends Central School and was especially thrilled when she met a former classmate at the book fair. "Thee is Agnes Morley, is thee not?" inquired Jeannette Leopold, and, in Agnes's mind, "a half century rolled away as though it had never been."

The Boston Book Fair (October 21–26) attracted even larger crowds than the Philadelphia fair. According to the *Boston Herald*, Agnes "stole the show" on the second day. The press described both her appearance and what had become a standard component of her talks.

> Sprightly and middle-aged, with greying, closely bobbed hair beneath a black hat, and a gay corsage brightening her black costume, she instantly set her audience of some 5000 book lovers a-titter as she stepped briskly up to the loud speaker and began to adjust it for the best effect. A few minutes later, the rafters were ringing with laughter and applause as she cupped her hands over a small harmonica, which she insisted on calling a "mouth organ," and in the best professional style tooted out a cowboy dancing tune as her final sally.

While in Boston, Agnes met with Houghton Mifflin editors and discussed ideas for a second book. They no doubt were pleased with the success of *No Life for a Lady.* It had been condensed in the October *Reader's Digest* and sales had reached 9,700 copies. The book also had appeared on best-seller lists in Boston, Los Angeles, and San Francisco. During her stay in the city, Agnes appeared as guest speaker at a number of luncheons sponsored by women's clubs. Despite the hectic pace of her schedule—speaking almost daily before one audience or another—she admitted that she was having "the time of her life."

Cleaveland's triumphal speaking and book-signing tour continued as she made her way home. On October 25, she autographed four hundred books at Kroch's Book Emporium in Chicago. She went on to New Mexico, where she spent ten days in Datil, mostly answering fan letters that had been pouring in since the book's publication. On November 6, she was one of five New Mexico authors who spoke at the Albuquerque Book Fair banquet, held at the Hilton Hotel. Pulitzer Prize–winning author Oliver La Farge was among the principal speakers. But, according to the local press: "In humorous approach, Mrs.

Cleaveland's address highlighted the program, and kept the audience of 150 persons in a continuous uproar of laughter." Julia Keleher, a writer and faculty member at the University of New Mexico, in her review of the fair, avowed: "Mrs. Cleaveland could outsell Dale Carnegie if she ever cared to capitalize on her technique in public speaking."

Agnes had numerous lecture engagements lined up after she returned to California. Foremost on her mind, however, must have been her upcoming appearance on the popular radio program, "We the People," scheduled for December 16. Reportedly, this was "one of the few radio hours that really sells books." But the attack on Pearl Harbor on December 7 changed everyone's life. In its aftermath, Agnes's appearance on the program was canceled—too late to stop her departure for New York City, where the program originated. She had flown from Los Angeles to Reno, Nevada, to make train connections for the East; only after reaching Chicago on December 13 did she learn of the program's cancellation. A telegram from Vivian Skinner of "We the People" explained: "Due to war we have been asked [to] present special morale programs and must therefore cancel our present schedule." Nevertheless, Agnes spoke on "We the People" on February 17, 1942.

Fan letters continued to arrive from all parts of the country. Some were from Agnes's former schoolmates; others were from Ray's classmates, all eager to share their stories about her remarkable brother. Typical were the sentiments expressed by Charles Cary, who knew Ray while both were students at the University of Michigan: "We called him 'Little Morley,' sometimes we called him worse things than that for his entry into the boarding house was always a signal for a riot of fun and amusement."

Many letters were from men and women who had worked or grown up on western ranches or lived in the Datil area and knew the Morley family. All agreed that Agnes had given an accurate account of range life. Millie Hamblet of Mina, Nevada, had read the book aloud to her husband, Sicle. Both had spent time in their youth on ranches, and both "loved" the book.

Sicle constantly interrupted the reading with remarks: "That's just the way it happened," "She sure has it all, just as it was—people and country," "That's it to a *T*." Charlie Rak, who had punched cows in the Datil area and knew both Agnes and Ray, wrote: "I want to tell you how good is your book. I read it from the first to the last in one sitting, and all the time I was watching for something to which I could say, 'Hell! That ain't so!' But not a word did I find."

Agnes heard from Lyle Vincent, the young cowboy who had been her mother's unofficial caretaker, and from Langford Johnston, one of Ray's Kindergarten Outfit. A letter from Bill Hazelwood, who makes an appearance in *No Life for a Lady*, confirmed one of her stories about riding "sidewise bareback on a horse" while trying to head off two mules running along a fence. After complimenting her on the book, he wrote: "The most unreasonible thing in your book for a stranger to believe was that bare back side wise riding you did after those mules. And I am ready at any time to sware that that is true."

She heard from people who had known her father and mother in Cimarron and from friends of Eugene Manlove Rhodes. Agnes long treasured the letters she received from May Rhodes, whose eyesight had so deteriorated that she hired a high school boy to read the book to her on three afternoons. "Agnes—you wonderful woman," May wrote. "Gene is so proud of you and of your book, and so are we all who know and love you."

Some readers referred to the outbreak of war and the uplifting nature of Agnes's book."You have done the nation as a whole a great good, letting the clean New Mexico breezes blow aside for a moment the smoke of battle," penned a woman from Santa Monica, California. Another writer from San José thanked Agnes for speaking to her friends on December 11, 1941. "Their spirits were wonderfully lifted 'in that awful week,' as many of them have told me."

After an edition of her memoirs was published in England in 1942, an appreciative reader there wrote to thank her "for the very real delight you have given me in reading your book, a welcome change from war absorptions." And a British officer,

Flight Lieutenant John C. Wilson, wrote from a German prison camp in September 1943: "We have just received your book 'No Life for a Lady' sent here from England. Personally I found it the best reading of any for a long time and we do a powerful lot of reading in this life."

No Life for a Lady was among the 1,322 titles chosen by a non-profit organization, called Editions for the Armed Services, Inc., to be mass produced as paperbacked, pocket-sized books and distributed to American soldiers. The books were great morale builders; fighting men carried them into foxholes and on dangerous bombing missions. One grateful serviceman, stationed "somewhere in Holland," wrote Agnes a poignant letter in December 1944, telling her that a buddy had given him a copy of her book, knowing that he had grown up on a ranch in New Mexico. "Little did I dream that in a few moments that book would take me right back home again and make me so darn homesick I pretty near cried." A *Reader's Digest* abridged version also found its way into a POW camp in the Philippines. A New Mexico prisoner, who survived the Bataan Death March, later credited Agnes's memoirs "with helping him through [his three-year] ordeal of horror."

Houghton Mifflin was eager to get a another book out of its prize-winning author, and Agnes was willing to comply. By early January 1942, she had settled on a topic: Ray Morley, a legendary figure in the Southwest cattle industry. Although she set about her task filled with enthusiasm, the writing did not come easily. Family problems and other distractions would delay the appearance of Cleaveland's second book by more than a decade.

CHAPTER 6

False Starts and *Satan's Paradise*

BUOYED by the success of *No Life for a Lady*, Agnes quickly set to work on a book about her brother, a colorful character whose exploits and shenanigans had become a part of the cattle industry's folklore. After publication of her memoirs, she had written a short piece about him entitled "Titan of the Range" at the request of George Fitzpatrick, editor of the *New Mexico Magazine*, who had long wanted to carry an article on Ray Morley. But this essay, which appeared in the December 1941 issue, barely touched on the many stories she had to tell about Ray—a cattleman, turned sheepman—who once dyed a small bunch of sheep red to explain to gullible tourists how Navajo women acquired naturally colored wool for their rugs.

In January 1942, Agnes wrote from Datil to her Houghton Mifflin editors, notifying them that a sample of the Morley manuscript would soon be in the mail. To which news, Paul Brooks responded, "I'll trust you to milk with one hand and write with the other—and both products will be Grade A." But this segment evidently did not meet Brooks's expectations. A kind and supportive editor—clearly captivated by Agnes's wit and vivacity—he couched his criticism of her manuscript in terms that she would find most helpful. "These pages do not yet suggest the book that we are both groping toward. . . . I do feel that you assume too much background on the part of the reader; you assume that he has already read *No Life for a Lady* and assume that he already knows something of Ray and that it is up to you to explain and justify him." He suggested that Agnes think of Ray "as typifying an era, a way of life, a pattern of thought. . . . if you can let yourself go on the old cowmen, you will have a book."

Ending on a positive note, he wrote: "You tell me that you are getting up steam, and that's enough for me. Once you're on the right track, there'll be no stopping you."

But distractions were many, for she continued to speak to civic groups and write an occasional essay on the state of world affairs. For example, in the fall of 1942 she sent to Houghton Mifflin a forty-eight-page treatise entitled "The Good Fight," in which she addressed the central questions of the day: "Why must we fight?" "For what do men die?" In keeping with his customary tact, Paul Brooks gently rejected her submission, saying "'The Good Fight' is full of good fighting words and says many things that need saying. But it's not, I'm sorry to say, the type of book on which H. M. Co. can go to town."

By the end of the year, Agnes had received good news from her London publishers: approximately 1,800 copies of *No Life for a Lady* had been sold in England, and the book had been well received by the critics. Paul Brooks was especially delighted with Elizabeth Bowen's review in the *Tatler* (London), in which the writer called it "a good book," filled with "anecdotes, incidents, alarms and adventures. . . . [Mrs. Cleaveland] still sees those distant days with a youthful clearness." In conveying his pleasure to Agnes in a post-Christmas letter, Brooks concluded it was "a far cry indeed from Datil to Bowen's Court."

Since returning from her book tour in the East, Agnes had spent most of her days in New Mexico. Her decision to live apart from Newton, who remained in California, was dictated primarily by Morley Cleaveland and the fact that "nobody appeared on the horizon" to take over her supervision. Besides, Morley seemed happiest staying at Howard Flat, and, according to Agnes, "goes into near-hysteria at any suggestion of leaving."

Then, too, Agnes felt a sense of freedom living in Datil, where she avoided Newton's critical remarks and was not "beholden to him for food or shelter or care of any sort." Living at Howard Flat was relatively cheap; she had plenty of firewood, as well as milk, butter, eggs, and beef to live on. She also had her own sources of income, from two annuities, which when combined brought in $75 per month, and from her book royalties, which, on sales to September 30, 1941, had amounted to $2,105. Still, financial matters were bound to generate friction between Agnes and her husband. The government's wartime ban on gold mining (she told a friend) "was a solar plexus blow to the Cleaveland source of income." Moreover, Newton had promised to pay the $50 per month interest on the Berkeley mortgage, as well as send Agnes $50 per month for Morley's care. When he reneged on both promises (and the interest on the mortgage was charged to Agnes's banking account in Berkeley), she blew up. She angrily wrote to Newton, telling him "to come out here and do the chores and let me write and support the family." She confided to her son Norman: "Just about one more straw and I'll sue for divorce on grounds of non-support and demand a division of the property."

Living on Howard Flat during the winter, however, was not easy, even for an experienced rancher like Agnes. On the property were two cows, one pony, one burro, some chickens, and a horse or two, all of which had to be fed and cared for. And the weather was often dreadful. On Christmas Day 1942, Agnes drove from Datil to Swingle Canyon in a near blizzard to deliver a telegram to her friend Marie Nourse, saying that her sister had died. On the way, she could not see much of anything, she recalled. "When I got into Swingle Canon things got really bad and suddenly I was in the ditch." She walked half a mile to deliver the telegram and then Mrs. Nourse "got a shovel and her chains and we walked back against the wind, slipping and sliding every step." Together, they maneuvered the station wagon back onto the road.

Later, she and Morley were snowbound for six days, unable to drive from the house to the highway. When the thermometer dropped to zero, the irrigation pipe froze, then busted, flooding the area between the house and the barn, "making a fine skating rink." She told Norman that his fleece-lined coat was "a lifesaver" when she did the chores. "I keep one glove on when I milk and use only one hand so I do not suffer. . . . [but] it takes all my time to keep things going, livestock, fires, meals, laundry, etc."

In the spring of 1943, Mary arrived at Howard Flat with her children, seven-year-old Beth, four-year-old Norman, and the newest member of the Wohlers family, two-year-old Mary Ann. Mary and her mother soon came to an agreement: Mary would oversee the "toy ranch" (the whimsical name bestowed on Howard Flat) while Agnes went to Santa Fe to concentrate on her writing. Before leaving, Agnes made certain that Howard Flat was fully equipped, a "going concern—wood, stock feed, garden planted up to the season's requirements, the hardest chores taken care of, everything that is humanly foreseeable arranged for." She also lined up Mrs. Nourse to come twice a week to help with the work.

And so, by April 14, Agnes was comfortably settled in the small apartment that her friends Jennie Avery and Grace Bowman, Santa Fe insurance and real estate agents, had offered her free of rent. She planned to write at least six hours a day, she told her son. She also reported on "a mysterious project" that was underway nearby—a project, as it turned out, that resulted in the world's first atomic bomb. "The Army has taken over about 100 square miles of land in the high mountain country and is building a city [Los Alamos], which report has it will contain at least 5000 people. All who go must agree not to leave the place for the duration except in extreme emergency and then under guard. No visitors allowed and all mail censored." She added: "Current joke is that it is a submarine base with an outlet through Carlsbad Caverns."

For several weeks Agnes enjoyed a respite from family worries. She wrote six hours a day and had a competent stenographer to make the final copy. After writing-hours, "something nice always happens," she reported to Norman, meaning dinner with friends or giving talks at women's clubs. Her peace of mind was broken in mid-May 1943, however, when she received word from one of Newton's associates that her husband had suffered a stroke, although it apparently left him with few adverse side effects. Still, Agnes took this as a sign that she had to take charge—to provide for Newton's future, as well as Morley's. She soon put in a bid to buy an adobe house in Santa Fe, which was large enough to shelter most of the family—a study for her and rooms for Newton, Morley, and Mary and her children, should Mary care to move in (Mary was then contemplating a divorce from her husband). "I am, or was, making headway on my book," she reported, "but this is pretty upsetting."

Disheartened when none of the other members of the family—except Norman—seemed to approve of her recent purchase of the Santa Fe house, Agnes put it on the market and by late summer had moved back to Datil. Nonetheless, she recognized that future developments depended on Newton's mental and physical condition. "He can't be deserted," Agnes avowed, "and it's my job to stand by." Possibly, he would take an interest in Datil, she surmised, for lately he had suggested buying a home in Socorro "for winter use and making Datil really homelike for nine months of the year."

Still, Newton continued living in Berkeley while Agnes managed affairs at Howard Flat. Whether she accomplished much writing is difficult to determine; she did, however, accept offers to speak before civic and other groups. In late August of 1943, she gave "personal reminiscences" at a Rotary Club meeting in Socorro, enlivening her talk with "Sandy Land" on the harmonica. And in November she returned to Santa Fe as houseguest of Grace Bowman and Jennie Avery and spoke to patients at Bruns General Hospital, the talk sponsored by the Business and Professional Women's Club.

Early in 1944, Agnes traveled to California, no doubt to check on Newton's condition, but also to take care of legal matters. Unbeknown to Agnes, Marion Jones Brinton, an old friend who had died recently, named Agnes in her will as sole beneficiary of her estate, which included a cottage and property in Carmel. She spent several days in this little seaside village, during which time she arranged to sell the property; she also gave a talk at the Village Book Shop on the writing of *No Life for a Lady*. Later, she told Norman that she expected to realize more than $10,000 from the Brinton estate, certainly a welcome boost to her finances.

In mid-May, after Agnes had returned to New Mexico, Newton's condition took "a turn for the worse." Norman, stationed in California with the U.S. Army Air Corps, hired a Russian woman for his father's nurse and promised to keep his mother informed of Newton's progress. Agnes soon headed to Berkeley, however, and most certainly was there when Newton suffered a fatal stroke. He died on August 22, 1944, at the age of seventy.

Agnes spent the remainder of the year in California and no doubt grieved over memories of the good times she had shared with Newton in their youth. But she also had to think of the future. It may have been while she was on the Pacific Coast that she sent a brief outline for a screen adaptation of *No Life for a Lady* to Allan Kenward, writer and playwright, employed by Metro-Goldwyn-Mayer Pictures. The son of one of Agnes's long-time friends, Kenward had written the play *Proof Through the Night*, a drama of nurses on Bataan, which had fair success on Broadway and was "a Hollywood smash hit" as the MGM film *Cry Havoc*, released in 1943.

In a letter dated January 30, 1945, Allan told Agnes what was wrong with her tale. It did not fit "the preconceived pattern" of a western; it lacked both a riveting love story and a tall dark hero. What MGM producers wanted was "a rootin'tootin' shootin' son-of-a-saddlesore who is half Robin Hood and half Rhet Butler." He then suggested a story that might catch the fancy of the producers: one that featured a love-hate relationship between a male and a female rancher, with the post–World War I depression in the cattle industry (as described in Agnes's book) serving as background.

On March 6, Allan again wrote to Agnes commenting upon a second synopsis she sent, one more in line with his suggestion to write a "shoot-em-up" cowboy story—but one having little to do with *No Life for a Lady*. Kenward liked the tale, but a producer at the studio did not. In his opinion, Cleaveland glorified the country instead of the storyline. What was needed was a beginning, middle, and end. And that seems to have been the end of Agnes's attempt to break into Hollywood.

Meanwhile, Agnes had returned to Datil in late January 1945 to prepare for a trip to Mexico. Her cousin Sylvanus Morley and his wife, Frances, had invited Agnes and her friends Jennie Avery and Grace Bowman to visit them at their hacienda in the outskirts of Mérida, the capital of Yucatán. The ABC outfit, as Frances called the trio, left on February 22 for New Orleans, where they boarded a Pan American clipper for the three-and-a-half-hour flight to Mérida. For Agnes, the entire trip was "an incredible experience." She and her friends spent three days at Chichén Itzá, the spectacular Maya ruins seventy-five miles southwest of Mérida, which Sylvanus Morley had begun excavating in 1924. A Cuban guide piloted them around in the daytime; their nights were spent at the Mayaland Lodge, where they were treated "like visiting royalty" because of their status as house guests of "Doctor Morley." A few days later, Sylvanus personally escorted them to Uxmal, Maya ruins about forty miles from Mérida, known for its "stupendous beauty."

One of the highlights of the trip for Agnes was getting to "really know" her cousin. They found out right away that they shared a family trait: "the love of argument, especially with another Morley." One dispute, she later recalled, "was precipitated when [Sylvanus] casually remarked that he would rather have written *Gentlemen Prefer Blondes* than have deciphered the Maya calendar. This thesis he debated with as much vigor as I used in denying it, while the bystanders stood ready to intercede with fire-extinguishers."

Upon her return from Yucatán, Agnes received welcome news from Ferris Greenslet of Houghton Mifflin, who reported that the sale of *No Life for a Lady* in the special Armed Forces Edition amounted to nearly 100,000 copies. "It was probably read and assuredly enjoyed by several times that many soldiers, sailors, marines, and aviators," he added.

There is nothing to indicate that Agnes made progress on her second book in the coming months—too many distractions continued to pull her away from her writing. Allan Kenward, having recently resigned from MGM, visited Agnes in the summer and became so enchanted that he wired his fiancée that "this countryside was the only adequately romantic setting in which to be married." Agnes helped arrange the nuptials, held early one morning on the Continental Divide, with Agnes the matron of honor. A breakfast at Howard Flat followed, after which Agnes set out with Allan and his bride, Torki, on a tour of regional points of interest—the James Hubbell ranch in Catron County for sheep-shearing, the Puye cliff dwellings north of Santa Fe, and the Baxter ranch at Tres Piedras—a place of remarkable beauty in Agnes's eyes. They witnessed San Juan Indian dancers on the plaza in Santa Fe and attended the Indian ceremonial at Gallup. The Kenwards long remembered Agnes's hospitality and found ways in later years to show their affection for her.

Even more demanding of Agnes's time that summer was the visit of young David Morley, another of Ray Morley's grandsons. This teenager had "so many ants in his pants that he has us all bedraggled," Agnes told her son. He possessed "a tremendous reservoir of undisciplined energy" and disrupted routine household activities with his endless questions. David seemed to enjoy his stay on the ranch, however, and asked to return the following year.

The summer ended with a wildly successful three-day rodeo in Datil to celebrate the rebuilding of the old Navajo Lodge, a task undertaken by Frances Martin, a local entrepreneur. After the death of Ray Morley, his daughter Faith and her husband, Les Reed, had taken over management of the lodge. Within a decade, however, the building had fallen into disrepair and had closed its doors to tourists and hunters. Faith Reed sold the property late in 1943, shortly before a fire burned this once-imposing edifice to the ground. Martin eventually acquired the site and began construction, although her cinder-block buildings had no resemblance to the old White House of Agnes's youth.

Martin treated the local folks to a free barbecue during the celebration, which attracted residents throughout Socorro and Catron counties. A pageant presented on Sunday, September 2, "told the story of the winning of the West," which the *Socorro*

Chieftain described in detail. Members of the cast included "Mrs. Agnes Morley Cleaveland, riding a burro and garbed in the spirit of the times," her daughter Loraine Keffer, astride a beautiful pinto, and David Morley, attired in a coonskin cap to represent Daniel Boone. The much admired Floripe Armijo (and friend of all the Morleys) was queen of the festivities.

Agnes spent a good deal of time during the summer and fall preparing her cabin for the coming winter. She had its walls lined with celotex panels and also installed a chemical toilet and a small circulating oil heater. She noticed a vast improvement over previous winters. In January 1946, amidst the worst snowstorm in memory, Agnes wrote, "We have had zero weather for weeks but the house has never gotten too cold to be liveable." The storm, which dropped eighteen inches of snow in Magdalena, kept Agnes and everyone else in Datil snowbound for some days.

Agnes also took stock of her finances that winter—with sobering results. She told Norman, "At present I am using all my monthly income and dipping into the principal to keep Howard Flat going. Too many cars, etc. . . . Had it not been for my royalties in *No Lady* and the Brinton estate I'd be asking my children for the price of a meal right now!"

Nonetheless, on January 30, 1946, she sent Mary (who had separated from her husband) a check for $500 to pay off the mortgage on her house and to help finance improvements. The money came from "the small balance in Pop's estate," she explained. Newton had willed all of his property to Agnes, with the charge to "share equitably" with the children as "she judges wise and expedient." Agnes wrote to Mary: "My judgement is to see that Morley does not become any greater liability to her brother and sisters than will be inevitable in any case and to see that your children are made as secure as possible." Thus, neither Norman nor Loraine would receive any monetary benefit from their father's estate, agreeing that it was to their best interest to insure Morley's continued support. From time to time, Agnes would send funds to Mary should emergencies arise and "will consider it a part of the 'share equitably' instructions." But, as she pointed out, "Pop's will was made when he supposed there would be something to share."

In the spring, Kay and Lovell Thompson of Houghton Mifflin descended upon Howard Flat to enjoy Agnes's hospitality and a tour similar to the one she had arranged for the Kenwards. They also discussed possible topics for Agnes's next tome. "They fell for New Mexico hard," Agnes reported to her son, and "tried to force an advance royalty on me for a new book. Of course I refused but promised I'd get busy."

The Thompsons' visit evidently provided the stimulus that Agnes needed to resume writing. In a letter to Kay dated May 8, she proposed writing a book about the town of Socorro and its picturesque inhabitants. Lovell responded at some length after discussing Agnes's ideas with Kay and Paul Brooks. They agreed that the topic would have to be broadened to be "the history of all southwestern towns." For his part, Lovell was partial to a subject that Agnes and the Thompsons had discussed while waiting for the train in Lamy—"Magic and Mystery of the Malpais," referring to lava flows in west-central New Mexico. He sent along a brief description of how he envisioned such a book would unfold. Choose the topic "that excites you most," he advised, "and let us see how it goes."

Several ideas, in fact, piqued Agnes's imagination. She soon sent Lovell an outline for a novel, featuring a strong woman protagonist who struggles to tame the land. She also continued to write about her brother, explaining in the preface to a manuscript entitled "I Remember Ray" that since the publication of *No Life for a Lady*, people had "bombarded" her with suggestions that she write more about her legendary brother. Another writer with an interest in Ray's history convinced her that if she didn't chronicle his life, "somebody else will, somebody like me—and then you'll be sorry." And so she turned to writing Ray's story, retelling the history of the Morley family in the first two chapters but with more amusing anecdotes than appear in her first book.

Lovell Thompson was delighted with the sample that Agnes submitted and thought that when completed "I Remember

Ray" would "make an excellent companion piece to *No Life for a Lady*." But he had not forgotten her proposal for a novel; he asked her to submit another rough chapter or two on both "your brainstorms." Then, he said, they could "advise as to which one we believe has the best chance."

While Agnes mulled over ideas for her second book, her essay, "Salt on the Tail of Yesterday," appeared in the summer 1946 issue of *Southwest Review*. With her good humor very much in evidence, Agnes described for her readers the difficulties encountered when she and her co-workers put together a rodeo to celebrate the reopening of the Navajo Lodge. They quickly found that old-time rodeos had become a thing of the past. Bucking horses were in short supply; not a single chuckwagon was to be found; nor any wild bulls to be ridden. "Nothing was *exactly* right." Still, Agnes concluded, "everybody came [and] had a bang-up good time."

Agnes Morley Cleaveland continued to enjoy good publicity throughout the year. She was featured in two articles by Thomas Ewing Dabney, editor of the *Soccoro Chieftain* and prize-winning journalist. The first, "No Life for a Lady," appeared in the June issue of the *New Mexico Stockman*, and was illustrated with pictures of Agnes on her ranch. "She still lives what is essentially a ranch life," Dabney wrote, "under the frontier conditions that make Catron county distinctive even in this day of surfaced highways and automobiles." In his second essay, "Background for a Book," published in the July issue of the *New Mexico Magazine*, he recounted in some detail how Agnes came to write her highly acclaimed book. Western author Frank M. King also wrote about Agnes and the Morleys in his privately printed *Pioneer Western Empire Builders, A True Story of the Men and Women of Pioneer Days* (1946).

Then, too, Agnes appeared before audiences in Albuquerque and Santa Fe during the summer and fall of that year. The most memorable of her talks was "Do's and Don'ts for Tenderfeet," presented before a standing-room-only crowd on the night of July 17 at the University of New Mexico summer session. "Laughter swept the audience of some three hundred repeatedly," one observer reported, "as she gave salty tips for tenderfeet, then afterwards spoke of her adventures as a best-seller author."

As winter approached, Norman urged his seventy-two-year-old mother to escape the harsh conditions in Datil and hunker down in Berkeley in relative comfort. Possibly Marie Nourse would care for Morley, he suggested. "The main thing you should be doing is writing. You could do that here if you wouldn't go crusading forth to save the world." She remained at the ranch. When she became bedridden with flu early in 1947, Norman chided her for ignoring his advice; he also expressed the opinion that she would never do much writing in Datil—too many distractions, he implied, too many opportunities to run away to Albuquerque or Santa Fe.

Agnes, however, continued to juggle both private and public events with aplomb. By mid-April 1947, she had written two political essays, for which she failed to find a publisher, and in May delivered two speeches in Albuquerque. Then the two oldest grandchildren, Beth and Norman, arrived in July to spend the rest of the summer on the ranch. And there were other visitors as well: the Morley cousins—brothers Griswold and Herbert Morley from California, and Sylvanus Morley's brother Henry from Virginia; also her old friend Charlie Rak; and two researchers from the Peabody Institute of Boston, whom Agnes guided to an old Indian cave in Thompson Canyon.

Agnes ended her social engagements that year with a talk at the School of Mines in Socorro. The local press gave her a fine buildup. "Cleaveland Roundup at Mines Friday," whooped the headline. "She will take you back into the old West, when men carried the law on their hip . . . when women rode side-saddle." On the evening of November 14, she delivered her address before a large crowd of students and townspeople, gathered in the school's gymnasium to be entertained by this nationally recognized author. Agnes did not disappoint them, for, according to the press, she made "the buckaroo days of the 1880's in the cattle West [come] to life."

Agnes's pursuit of a second book temporarily came to a halt early in 1948 when she set out on one of the greatest adventures of her life—a six-months' trip to Malaya to visit her son in Kuala Lumpur, where he was manager of the Pacific Tin Consolidated Corporation. A bachelor and a thoughtful son, Norman probably financed most of the trip, for he often sent funds to cover his mother's expenses. Family friend Emily Dole, formerly of Hawaii, now living in Riverside, California, stayed with Morley during Agnes's absence.

Agnes sailed from San Francisco on January 24, on the Norwegian freighter *Bougainvillea*. Her traveling companion was an old friend, the New York magazine writer Edna Nelson. Friends of both women crowded the ship's passageways prior to departure to bid them goodbye. In Agnes's stateroom were "letters, telegrams, and a dozen magnificent roses from Kay and Lovell Thompson—just to prick my conscience," she quipped.

The *Bougainvillea* docked in Los Angeles after dark the second day out to take on cargo. On the 26th, Agnes's daughter Loraine and Allan and Torki Kenward came on board and joined Agnes for lunch at the captain's table. The Kenwards gave Agnes a large journal "with the hope," they told her, "that words would deface its pristine pages." Indeed, Agnes filled the journal with detailed descriptions of her journey, which took her to Manila, Shanghai, Hong Kong, and Singapore. During the voyage, she incorporated her journal entries into a series of letters she mailed to the *Socorro Chieftain*, where they were published.

Delighted with the ship's accommodations early on, Agnes recorded, "private bath, luxurious stateroom"; days later she reconsidered, "plumbing fixtures—no good." The ship's crew, young Scandinavians, soon erected a swimming pool on deck, which Agnes described after testing the waters as "a sort of glorified bath tub—three in at one time is a crowd." To Agnes's great pleasure, the crew assembled on deck at night to sing, accompanied by a violin; a favorite tune was "Home Sweet Home," although, she confessed, "I can't understand any of the words, if they are intended to be English or not."

Her fellow passengers included two Norwegians, two Dutch, one Chilean, two Americans (in addition to Agnes and Edna), and a family of three from Canada. At an opportune time, she circulated two copies of *No Life for a Lady* and gave an autographed copy to the captain. Mrs. H. of Norway "asks a lot of questions I find hard to answer especially about the 'ronch'[ranch], who milks all the cows, what we eat and all manner of details. I feel she doesn't get the picture at all although she is reading No Lady and professes to enjoy it."

The passengers entertained themselves for nineteen "cloudless days during which we saw no ship, plane or sign of life other than a whale, between Los Angeles and Manila Bay." Finally, they threaded their way "past tragic Corregidor and Bataan, through a maze of sunken battleship masts, and docked at what was left of a once fine pier." Two and a half years after the close of World War II, Agnes described Manila as "a ghastly pile of ruins; public buildings, cathedrals and palatial homes are rubble heaps." Most surprising was the language that Filipino youngsters had picked up from American GIs: "Every American man is called Joe, and American woman Baby. It was a little startling to me when I was hailed on the streets by smiling urchins, 'Hello, Baby.'"

On the voyage north to Shanghai, the climate changed. "It is much colder, but the sea is not rough,"Agnes wrote in her journal. "I've resorted to Norman's wool slacks, the sweat shirt George [Keffer] gave me and a sweater, but even so can't stay out on deck very long." When she went ashore in Shanghai, she was nearly overwhelmed—although she had been forewarned. "People, people, people, the streets pack jammed[,] street cars, autos[,] jeeps[,] rickshaws, bicycles, moving in the midst of a solid mass of pedestrians."

From Shanghai they sailed south to Hong Kong and then to Singapore, where Norman was on hand to greet his visitors. Then they flew to Kuala Lumpur "over 200 miles of forest and rolling hills, much of it rubber plantations." Once ensconced in Norman's quarters, Agnes and Edna were assigned their own "Chinese handmaiden." Agnes did not quite know what to make of the British colonial system, which provided far too many servants, who, in her opinion, were much too helpful.

But she followed local custom, including the obligatory afternoon siesta. She also attended the Governor's annual garden party, along with five hundred other well-dressed guests. "It was a colorful occasion," she recorded. "Dignitaries of many races. Chinese, Indian and Malayan in costume of bright silks and jewels gave an air of a Hollywood stage set."

Toward the end of Agnes's sojourn, a series of strikes erupted in the coastal cities of Malaya. As the strikers became more militant and violence broke out, the British colonial government declared a state of emergency, which was not lifted until1960. During the "Malayan Emergency," as the British called it, the government enacted measures to suppress "left wing political movements," including the Malaya Communist Party. Agnes described some of her own anxious moments: "The last few days of my stay in Kuala Lumpur were tense because my son Norman had been deputized as an inspector of police with a considerable force under him and told never to go abroad without being armed. Some of my sightseeing plans had to be abandoned as too dangerous."

Agnes sailed for home on June 27 aboard one of the American Mail Line ships and in less than a month was back on American soil. During her absence, a German edition of *No Life for a Lady* had been published, under the title *Im Lande der offenen Weiden* (In the Country of the Open Pastures). Her *American Primer* had also attracted notice when, on May 11, Congressman Gordon L. McDonough of California inserted it in the *Congressional Record*, with the recommendation that it be read by every member of Congress. After that, Garet Garrett, editor of *American Affairs*, contacted Agnes and secured her permission to republish it as a supplement to his journal. "Not an extensive audience perhaps but a discerning one," Agnes wrote to Mary. "I'm quite pleased."

For the remainder of the year, Agnes worked sporadically at her writing. She evidently sent to Houghton Mifflin a revised essay that she had first submitted prior to her trip to Malaya, entitled "I Go Alamo," a treatise on the Alamo Navajo Indians. (The firm turned it down because of its brevity.) And following the death of her cousin Sylvanus in September, she wrote a humorous piece about her trip to Yucatán, "Prima Gringa Discovers a Relative," to be published in *Morleyana: A Collection of Writings in Memoriam Sylvanus Griswold Morley* (1950).

Sometime following her return from Malaya, Agnes made contact with Fred Lambert, a former law officer, poet, and artist—and the motivating force behind her second book, *Satan's Paradise*. Like Agnes, Fred Lambert had been born in Cimarron, New Mexico (in 1887). Sworn in as a deputy sheriff at the age of sixteen, he spent the next thirty-five years as a New Mexico lawman. Now retired and living with his wife, Katie, in Kansas City, Missouri, he had sent Agnes a copy of his recently published *Bygone Days of the Old West* (1948), a collection of his cowboy poems and pen-and-ink drawings. In April 1949, he asked Agnes to write a foreword for a new book—his memoirs of the Cimarron country. But, after he heard "indirectly" that Conrad Richter was urging Agnes to write her own book about Cimarron, Fred proposed that they collaborate, "strike up a 50-50 contract," and publish the volume together.

In a series of letters, starting June 25, 1949, Fred outlined his plan. He had all the stories written, he told Agnes, but she was to feel free to add some of her own. He confessed that he was no writer. "I know definitely that you could take my Cimarron stories and any other data that I could furnish, and with anything you might add to them and we would then have a book worth the reading." In September, Lambert flew to New Mexico to consult with Agnes. Evidently, she did a good deal of background checking on Lambert and his stories, which did not upset Fred in the least. Rather, he and Katie were pleased that Agnes seemed to have "the 'blue prints' all made up for a wonderful book." After further correspondence, Cleaveland and Lambert signed an agreement in October to share equally any "monetary profit" accruing from their book, which "neither of us alone could produce."

In short order, Agnes sent a partial Cleaveland-Lambert manuscript to Houghton Mifflin. Paul Brooks's response on March 31, 1950, indicated that her sample material did not yet merit a book contract. He called "the whole project . . . terribly provocative" and Fred Lambert's material, "unique and valuable." But, he continued, "the process of making this material into a book is another matter. What we have now is scattered bits, quite fascinating in themselves and with some narrative quality, but still disparate." Kay Thompson, who was then in London, possibly could give the manuscript what was needed: a beginning, middle, and end. Brooks passed along Lovell's suggestion that Agnes come to Boston after she had a manuscript of 100,000 words. In returning the sample, Brooks offered Agnes a modicum of hope: "We are willing to work hard to quarry out of this a successor of *No Life for a Lady*, but I don't see it quite yet." Evidently on the basis of this remark, Agnes informed Mary—stretching the truth a bit—that Houghton Mifflin had accepted her new book.

In June, and again in December, Agnes flew to Kansas City to discuss matters with Fred. Meanwhile, her granddaughter Beth Wohlers arrived in July to spend the rest of the summer with Agnes and stayed on in the fall to enroll in the Magdalena high school. By this time, Agnes's eyesight had diminished to the point of near blindness. This was due, she wrote to Mary, to inoperable cataracts, which left her unable to sew or cook. "I still read if conditions are exactly right and manage to scrawl with a pen, as you notice!" She added: "Beth is always willing to relieve me of little tasks which are difficult for me."

Mentally, Agnes seems to have come to terms with her limited vision. In another letter to Mary, Agnes recalled that in her youth, she (Agnes) liked to wander around after dark, doing tasks she could have done in the daylight. "I still like to grope in the dark. It's a phase of adventure and maybe one reason I am not terrified at any threat of going blind."

Despite Beth's helpfulness, having a teenager in the house was not always easy. Norman had provided Beth with a jeep to drive from Howard Flat to Datil to catch the school bus into Magdalena. The bus returned to Datil about 5:15 P.M. On a night in mid-January 1951, when "it was snowing and blowing a gale" and Beth was late in getting home, Agnes worried that she had become stuck in a snowdrift. Taking Morley along "to be my eyes," she drove to Pine Park—the nearest place having a phone—to call to see if the school bus had arrived. "It was a fool hardy stunt on my part," she confessed, "because I haven't vision enough to drive at night even under the best condition. . . . When I got to Pine Park after a rather terrifying experience I knew I could not drive home again. It was snowing and blowing harder and the snow would pile up on the windshield." All turned out well, however. A phone call confirmed that the school bus had arrived on time; and Carl Simpson, a local resident, helped her drive home, where she found Beth, safe and no doubt contrite that she had caused her grandmother so much worry.

Not long after this escapade, Agnes sent a revised Cimarron manuscript to Houghton Mifflin. In March 1951 she received two reports, one from Paul Brooks and the other from Kay Thompson. Fortunately for Agnes, Kay's more optimistic letter probably arrived a day ahead of Paul's disheartening missive. Although admitting that Agnes's manuscript could not be another *No Life for a Lady*, Kay was willing to work with Agnes to shape it into a book worthy of the Houghton Mifflin imprint. "The book needs cutting and rearranging and simplifying—you know, the old routine. Would you be willing to let me have the manuscript for a month and see what I could do—at no expense or commitment to either you or HMCo?"

Paul Brooks, on the other hand, bluntly told Agnes that her manuscript was not publishable. "It has a great deal of interesting material," he conceded, "but it hasn't the pattern nor (and this is even more important) the color and vitality and sharp edge of *No Life for a Lady*. It hasn't, in short, the personality which came through on every page of that book." Still, he was willing for Kay Thompson "to take a whack" at improving the manuscript. "If she wants to try, God bless her."

Kay set right to work. She obtained books that described Cimarron's wild past to check details in Agnes's manuscript; she fired off questions to both Fred and Agnes to clarify knotty issues; and she made a hurried trip to Howard Flat in May to consult with the author. As Kay neared the end of performing her editorial wizardry, Paul Brooks conveyed to Agnes his revised opinion of her manuscript, now entitled "Cimarron Means Wild." "I think that it has shaped up remarkably well under [Kay's] guidance. . . . Though I never expect to feel about a Western book quite as I did about *No Life for a Lady*, I think this is a worthy successor." Although Kay still had more work to do on the manuscript, the firm issued a contract to Agnes and Fred Lambert on November 23, 1951, the book royalties to be divided equally between them.

In the midst of her editorial work with Kay Thompson, Agnes engaged in what seems to have been a favorite activity—entertaining young people on Howard Flat. In July of 1951, two separate groups of Girl Scouts camped on her property while taking part in Bertha Dutton's Girl Scout Mobile Archeological Expeditions. An archeologist, ethnologist, and friend of Agnes's daughter Loraine, Dutton had begun the Girl Scout Mobile Archaeological Expeditions in 1947. Each summer she led groups of teenage scouts (from all parts of the country) on camping tours of significant historical and archeological sites in New Mexico and Arizona. Over the years, the routes changed to provide new experiences for returning campers. The brochure describing the 1951 tour lists Datil as the fourth stop: "Datil, made famous by pioneers, of whom Agnes Morley Cleaveland has written such a fascinating account in 'No Life for a Lady.' *Read it*, for you will spend a night on her ranch, and hear tales omitted from her book." Agnes undoubtedly enjoyed talking to the scouts as much as they enjoyed their two-week tour.

The rest of Agnes's year was filled with civic events and entertaining guests at Howard Flat. As winter approached, she fielded more questions from Kay Thompson. In a Christmas letter, Kay revealed much about her role as editor when she called Agnes's attention to the beginning paragraph of Chapter One: "[This] smells of me and not of you, but as you know this is for you to change where you don't like it, and represents transitions between hunks of AMC. It is I think the only paragraph in the book that is so manhandled, and does not represent a trend."

Agnes flew to Boston in January 1952 at the request of her publishers. Beth accompanied her grandmother, for, as reported in the *Socorro Chieftain*, "Mrs. Cleaveland is no longer able to see to read, and getting about is a very dangerous matter with traffic as it is." They were gone ten days, apparently stopping over in Kansas City to visit with the Lamberts on their way home.

When Agnes's book appeared in the fall of 1952, its title had been changed to *Satan's Paradise, From Lucien Maxwell to Fred Lambert*. Cleaveland is listed as the sole author; Lambert, as creator of the drawings that accompany the book. That Houghton Mifflin failed to acknowledge Lambert on the title page as Agnes's collaborator is a puzzlement. Possibly the publishers believed that Cleaveland's name alone would sell the most books. At any rate, Agnes acknowledged in the book's opening pages that, although she had done the writing, Fred "has done the remembering." She also avowed that his vast collection of documents, newspaper clippings, and other memorabilia had been used to substantiate the book's accuracy.

The twenty-three chapters that make up *Satan's Paradise* tell the story of the birth of Cimarron in the late 1860s and its turbulent history thereafter, as lawmen like Fred Lambert worked to tame the wildness of the town and the region. In lively prose, Agnes relates tales about the early giants in that history—Lucien Maxwell, Uncle Dick Wootton, Henry Lambert (Fred's father), and her own father, William Raymond Morley. She also includes engrossing profiles of those who fell afoul of the law—Clay Allison, Charles Kennedy (a brutal killer and wife abuser), and Black Jack Ketchum.

The heart of the book concerns Fred Lambert and his adventures as a lawman, which started at age sixteen, when he single-handedly captured three alleged killers. Drawing on Fred's memories, Agnes weaves together thrilling and amusing accounts of his tussles with lawbreakers in this country and in Mexico. A few of the stories involve Katie Lambert, who often accompanied her husband on official duties. Fred Lambert emerges from these pages as a wise and gentle law officer, one who preferred using his wits rather than his six-gun to apprehend miscreants.

Within weeks of the book's release, Lovell Thompson informed Agnes that sales had reached four thousand copies. "We think old Satan has had a pretty good reception," he wrote, "and we are well pleased to find ourselves in partnership with him." Indeed, many reviewers had nothing but praise for the book. A. B. Schuster, in the *Oakland Tribune*, called it "a thrilling and informative book. . . . Our Mrs. Cleaveland has done it again." Floyd (Slats) Logan of the *Fort Wayne News-Sentinel*, called Agnes "a born story-teller." "Her style is vivid," he continued, "her choice of words is distinctly impelling and her discernment is arresting and exciting."

Inevitably, some critics found *Satan's Paradise* less compelling than *No Life for a Lady*. J. Frank Dobie, in the *New York Times*, judged that Cleaveland's second book was not the peer of her first; but, he concluded, "'Satan's Paradise' is, nevertheless, lively reading." The sharpest criticism came from Oliver La Farge in the *New York Tribune*. Much of the book, he stated, lacked "the zip and realism of 'No Life for a Lady.'" He also criticized at some length the author's misuse and misspelling of Spanish words, attributing this carelessness to "an element of racial feeling"—meaning that the speech of "the lowly Spanish-Americans and of the Mexicans across the border" was "not worthy of an effort towards correctness." *Satan's Paradise*, he added, was "full of the old, true, untamed Southwest," but the presentation was "not worthy of the material."

La Farge's "blistering attack" upon her book and her integrity wounded Agnes deeply. Because of her impaired vision (she later explained), she could not read the galley sheets of *Satan's Paradise* to correct the spelling of Spanish words. Houghton Mifflin was supposed to do that for her. In fact, the firm responded to La Farge's criticism in an open letter to the *Tribune*. But, in Agnes's words, "apology and explanation can never over take an unfair accusation."

J. Frank Dobie and Oliver La Farge were both partly right. Agnes's stories in *Satan's Paradise* make for lively reading, but they lack the personality that permeates the pages of *No Life for a Lady*. Agnes's second book seemingly never captured her imagination. Probably she would have been better served had she carried through with "I Remember Ray," a story she could write from the heart, giving her wit and creativity full reign to bring Ray Morley to life.

Still, after publication of *Satan's Paradise*, Agnes must have enjoyed being in the spotlight again. There were book signings, fan letters from as far away as Japan and the Netherlands, and family members rallying to her support. Norman wrote from Malaya: "A nice letter from Aunty [Lora Reynolds] reports that you are receiving much acclaim for the new book and that only La Farge has been a stinker."

By the time her book was published, Agnes's eyesight had deteriorated to such an extent that she hired local residents to help on the ranch and enlisted others to read to her. A steady stream of visitors enlivened her days. She continued to travel about the countryside, volunteers driving her to Albuquerque (to give a talk on Billy the Kid) and to Santa Fe (for a Thanksgiving dinner with her daughter Loraine and her second husband, George Lavender). In mid-December, she met Loraine and Jennie Avery in Albuquerque, where they boarded the Super Chief for Los Angeles to attend a conference of Christian Scientists. As the year 1952 came to a close, the seventy-eight-year-old Agnes Morley Cleaveland—despite her failing eyesight—seemed just as feisty as she ever was.

CHAPTER 7

"Post Mortem"

NORMAN Cleaveland once told a reporter: "[My mother] loved being on a horse, and she was daring and independent and rebellious." This description holds true even into Agnes's final years. She always wanted to be active, to do something exciting, to make the most of her days. She also wanted to connect with other people—to entertain them and make them laugh, much as she had done in the past.

Moreover, Agnes never lost her passion for writing. Even though legally blind, she found she could write letters and record her thoughts in spiral notebooks (as one neighbor described it) by "keeping the left finger on the margin, writing a line with the right hand, then moving the left finger down about half an inch each time," although sometimes her sentences trailed off diagonally across the page. Among her unpublished manuscripts is a description of a scientific project conducted by Katherine H. Clisby of Oberlin College on the San Augustin Plains during the summer of 1952. She and a colleague spent a weekend or more at the Cleaveland ranch before finalizing plans to drill several hundred feet in the plains (an ancient lake bed) to test "soil for former vegetation" in hopes of determining climatic change over the years. One day Dr. Clisby took Agnes on a jeep tour of the area the scientists meant to test. "It was no tax on my mind's eye," Agnes wrote, "to see a forest carpeting the rolling beaches, bars, stretching up to the extinct volcanos along the rim of the lake bed."

Agnes also autographed copies of *Satan's Paradise* (probably with assistance). At a book-signing party at the Villagra bookstore in Santa Fe, she quipped: "You can't say I'm a fast writer."

It had been eleven years since she last sat in a local bookshop to sign copies of *No Life for a Lady*. "I guess the next will be ready in another 11 years," she added.

Agnes carried on in 1953 as if being blind were no handicap at all. She continued to be a popular lecturer. She gave talks to Farm Bureau meetings in Socorro and in Datil, to a summer session at Highlands University in Las Vegas, and to townspeople and students at New Mexico Western College in Silver City, where a dinner in her honor was held in the college dining hall.

She rejoiced with other Datil residents when they finally received telephone service that summer. Work started in the spring, with local men helping when they could; Agnes loaned her jeep for the wire work. Young Gary Tietjen, who had moved with his family to the Drag A Ranch in 1949, later recalled that Agnes contributed her labor as well. Each rancher was responsible for building the line from the highway to his or her ranch, he explained. Post holes had to be three feet deep, but posthole diggers could only remove about two feet of dirt. The last foot of dirt had to be removed with a one-pound coffee can. And there was Agnes, "lying on her stomach out there with her coffee can, scooping that last foot of dirt out. She loved it."

Agnes enjoyed Gary's company. He often read mail and books to her, helped her to write checks, and on occasion served as her chauffeur. In September they drove to Socorro to support a move to establish a commercial bus line connecting Quemado and Socorro, with stops in Datil and other small communities along the way. When the daily service began on October 19, Agnes held ticket number one and made the round trip with four other passengers.

One day Agnes begged Gary to take her on a hike up Thompson Canyon. "She was blind," Tietjen recalled, "but she could walk quite well with the aid of a stout stick which she called a *bordon*. She needed someone directly in front of her to follow. She could barely make them out. She wanted to go up some rather steep hills and we did. When we came back down, she would sit down and slide. This wore her pants out. She didn't want anyone to know that, so she made sure I did not get behind her."

Gregarious since her childhood, Agnes enjoyed entertaining family and friends at the ranch. She probably was happiest when Norman flew in from Malaya, which he did twice during 1953. Loraine seems to have been a frequent visitor, and on at least one occasion Mary drove down from Montana. Other summer visitors included George Stevens, the son of Montague Stevens (of grizzly bear fame), and Lelia McPherson, the widow of Orrin McPherson, Ada Morley's cousin. In the fall, Agnes welcomed a Methodist youth group to her picnic grounds among the pines and also staged an outdoor party for a large gathering of adults and youngsters.

But Agnes was no longer capable of spending winters at Howard Flat. In November she went to stay with Loraine in Santa Fe. She returned briefly to host a Christmas party in Datil for some eighty guests—a grand success, according to one attendee. "Each and all were given presents of sorts. Jokes and funny things were among the gifts."

In a similar fashion, Agnes kept active in 1954. Gary Tietjen remembers that she could still see well enough "to make her way around her place pretty well." He also recalls that he was reading a book to her on space travel—a topic that fascinated her—when he was drafted into the army and the book was never finished. Knowing that he was a Latter-day Saint, Agnes sent him a letter: "Dear Saint Gary, you abandoned me up here on the moon. Send troops to get me off of here. Agnes." Gary never saw Agnes again but remembers her as being "spry and active and funny."

Agnes again hosted Girl Scouts on her property that summer, telling them "tales of the Old West and how she came to write 'No Life for a Lady.'" But the project that absorbed most of her attention was the relocation and renovation of the old Jack Howard homestead cabin. A new well had been dug a mile or so from the main house on Howard Flat; Agnes then had an old wooden windmill installed—something to show her guests, she said. Next she moved the cabin to the site, which she now referred to as "Cliffden," and hired local workers to make the domicile "modern and attractive." Yet during the following year, she reached back into the past to create a living history museum. She acquired a wood-burning stove from George Stevens (she refused to have an electric one); and Norman sent her a 1916 Model T Ford. The day after the vintage Ford arrived, Hillery Summers, who had headed the Cliffden project, drove Agnes into the village for picture taking, he dressed in an old-style costume with a sweeping moustache, duster, and goggles, and she sporting a wide hat, tied down with an old-style veil. She planned to give local schoolchildren rides in "Mistress Lizzy."

Callers seemingly enjoyed their visits to the Cliffden Ranchito, as some called it; others referred to it as Agnes's doll house. On the premises was a baby burro, the gift of a local resident. Agnes also had on display many novelties that she had collected over the years. Neighbors came to clean and cook for her; and old friends from Santa Fe and Albuquerque also brought food and read to her.

But it proved difficult to find attendants who would stay full-time at Cliffden. Consequently, when Mary Wohlers arrived with her fourteen-year-old daughter Mary Ann to spend Thanksgiving with Agnes in 1955, she was shocked by her mother's condition. Agnes apparently had to fend for herself some of the time, telling Mary that she ate raw eggs "of necessity"—she couldn't see to cook on the wood-burning stove. Despite the generosity of neighbors and friends who delivered food, Agnes had become alarmingly thin.

Understandably, Agnes had lost some of her buoyancy as well. After all, she could no longer move freely around Howard Flat. She could no longer drive. She had to depend upon others to get her groceries and her mail. To a person like Agnes, whose life centered on reading and writing and constant activity, impaired vision at times must have led to despair.

Mary immediately wrote to Agnes's long-time friend Dr. Evelyn Frisbie, an Albuquerque physician, expressing her fears about her mother's condition. Frisbie, who had been to see

Agnes in August with Jennie Avery, assured Mary that "the situation is not quite [as] serious as it must have appeared to you," although she agreed that Agnes looked severely undernourished. Despite a steady stream of callers, what Agnes lacked most was companionship. "When we arrived," Frisbie wrote, "she was sitting alone staring into space—expecting no one as everyone had gone to the rodeo and she had no hope of visitors. With her temperament, she is very dependent on communication of ideas, mental companionship and praise and adoration of intellectual accomplishments." Still, following Mary's visit, conditions improved. Agnes acquired a full time housekeeper, "which must make all the difference in the world both physically and mentally."

Agnes also had resumed writing, which, Frisbie avowed, "must be the best possible outlet for her." Starting in January 1956, Agnes began a second book of memoirs entitled "Post Mortem." In the opening pages, she tells of colliding with an old-timer (possibly in a Magdalena store), who blurted: "Why, Mrs. Cleaveland. I heared you was dead." "I am," she countered. "But I'm trying to keep it a secret. I died last week. Promise you won't tell anybody." Muttering as she made her way out the door, she left her attendant wondering what she was talking about. The driver of the car remained silent. "It was an expressive silence," Agnes writes, "which said 'She is losing her mind as well as her eyesight. Poor old thing! It must be awful to go blind at 80.'"

Agnes then tells of the day when she realized she had died. "It was when I walked into the room of a lifelong friend prepared to hear her usual 'Oh, here you are. Sit down and tell us all the latest news about that incredible Datil of yours.' What I heard, instead, was 'Sit right here, my dear,' spoken in a stage whisper. 'We are listening to the afternoon radio and television program. We just bought a 21 inch TV set.' And I knew that never again would my news about Datil or anything else be of the slightest interest to anybody. The world I had known was gone forever. What else is death? But I was still myself wasn't I? Life was still life wasn't it?"

And so she decided to record more of her memories about the Datils. Over the next several months, she struggled to describe in three spiral notebooks a series of episodes in her life, numbering them Memory No. 1, Memory No. 2, and so on. The handwriting in the first notebook is sufficiently clear so that about two-thirds of it has been transcribed. But midway through the second notebook the writing progressively deteriorates until it becomes illegible. Still, several episodes in "Post Mortem" have been incorporated in this biography and will live on as part of Agnes Morley Cleaveland's legacy.

In the spring of 1956, Agnes moved to Orinda, California (near Berkeley), to live with Mary and Mary Ann in a house that Norman had purchased for his mother. Agnes's children probably persuaded her to leave Cliffden, for in the fall she wrote to a friend: "As you have evidently heard, I was whisked off to California by George and Loraine Lavender and have not, as yet, been able to get back to New Mexico where my heart really is." She remained in California a little more than a year, much of the time in poor health. Mary Ann Montague later recalled that her grandmother "rarely was able to be out of bed. . . . She had the 'master bedroom' that was light and airy. She liked to just listen to the world, birds and all." She also used a "writing board" to correspond. Mary Ann described this as "an ingenious arrangement with a sliding plastic bar to keep her from writing over the last sentence above."

Norman Cleaveland hired a companion to read to his mother several times a week. On one of her better days, she sat outside chatting with Herbert Eustace, her longtime Christian Science mentor. According to Mary Ann, "talking was always a favorite pastime." Indeed, Agnes enjoyed talking with friends on the telephone; she also enjoyed listening to the radio and to the television set that she had purchased for the household. One day, however, she fell and broke a hip, which required a hospital stay. Norman flew back from Malaya to check on her.

As her health continued to decline, Agnes longed to return to New Mexico—vividly recalling how her brother had died in California "so far from his roots." According to Norman,

"[Agnes] coerced her West Coast physicians into letting her board a train," in June 1957. Quite ill when she reached Albuquerque, she was taken to Presbyterian Hospital, where Evelyn Frisbie awaited her. She spent several months in the hospital before she convinced Norman to take her to Datil. On New Year's Day, Agnes returned to Howard Flat in an ambulance accompanied by two nurses. She died at home of pneumonia on March 8, 1958, at the age of eighty-three.

Private funeral services were held in Albuquerque; her ashes were placed atop Mount Fuji (so named by Agnes's daughter Loraine), a mountain near Datil "facing the canyons and meadows she loved so well." An obituary in the *Albuquerque Journal* focused on Agnes's life in New Mexico: punching cattle, hunting grizzly bears, "imparting the feel of the West into her writing of short stories and books." The obituary that appeared in the *Oakland Tribune* highlighted her work in California—as a prominent Republican, a leader in the California Federation of Women's Clubs, the California Writers' Club, the Berkeley Political Science Club, and as president of Pro America.

Agnes Morley Cleaveland will long be remembered for the coming-of-age stories she told in *No Life for a Lady*. Her skills as a writer, however, must have been apparent during her early years when she turned in school essays about life on the ranch. Her career as a writer began shortly after her marriage to Newton Cleaveland, when she wrote dozens of short stories that appeared in some of the nation's most popular magazines. Her output truly was impressive. Like Caroline Lockhart, a western author three years her senior, Agnes based many of her tales on personal experiences. And like Lockhart's, Cleaveland's stories often featured a strong female protagonist.

It was during these early years that Cleaveland helped found the California Writers' Club, a club that continues to exist today. Not long after her death, the club paid tribute to Cleaveland's memory as a pioneer and writer in a special ceremony held at the Woodminster Amphitheater in Oakland. At the conclusion of the event, a plaque bearing her name was placed on the California Writers' Club Tree. Today club members look upon Cleaveland as one of their distinguished forebears.

Agnes gained material for her stories from a first-hand acquaintance with ranch life. Those early years on the ranch also helped mold her personality. Like many other ranch children, Agnes developed into a confident, self-reliant, and high-spirited youngster; and these character traits carried over into adulthood. Moreover, the picture Agnes paints of ranch life in *No Life for a Lady* finds validation in later accounts of women ranchers. Teresa Jordan, in *Cowgirls: Women of the American West* (1982), for example, focuses on several strong and independent ranchwomen, who, like Agnes, were deeply attached to the land, loved to ride horses, were good with animals (as well as good storytellers), and were treated as equals by male cowhands.

From her childhood on, Agnes Morley Cleaveland was determined to make her own way. A good example of this was her decision to move to Silver City (in 1907) and run the Rosedale Dairy rather than join Newton in California. Her work as a Christian Science practitioner can also be construed as an effort to find a life outside the family—and to make a difference in the world.

Still, the propensity of both Cleavelands to travel, Newton on business trips, Agnes on jaunts to the Datils, Europe, and elsewhere, undoubtedly put strains on the marriage. Indeed, tensions within the family may have been a major reason for Agnes's prolonged stays in New Mexico. Agnes's great-niece Nancy Reed Miller (the granddaughter of Ray Morley) recently offered the opinion that none of the Morley siblings—Agnes, Ray, Lora—were marriage material because of their strong egos and independent spirits.

Once her children were grown, Agnes immersed herself in club work and political activism. Women's clubs had proliferated in the first three decades of the twentieth century. During the 1930s, Agnes was among the tens of thousands of American women who worked for social and political reform. She held powerful convictions and invested much emotional energy into her political and club work. Very much a public woman, she was not afraid to voice her opinions openly on the chief issues of the day. Ambitious, as well as a talented speaker—always at ease before large audiences—she made a dynamic presiding officer in the many clubs she joined. And she was remarkably effective as a grassroots political activist, earning the respect of such well-known political figures as Herbert Hoover and Earl Warren. Her friend Lou Henry Hoover also enlisted Agnes's aid in advancing her program to strengthen traditional American values. That Agnes recovered so quickly from her setback as president of Pro America underscores the resiliency that was a major part of her character.

What seemed special about Agnes Morley Cleaveland's personality was a certain magnetism—or call it dynamism—that attracted a host of devoted friends, young and old. Nancy Miller recalled the time in her childhood when, after her mother (Faith) told her to put on a coat before going outside, Agnes jumped to her defense and declared that the seven-year-old was old enough to know for herself when she was cold. From that moment on, Nancy avowed, "Aunt Agnes was my hero." And she added, "Everyone thought that Aunt Agnes was pretty much all-knowing about everything. That was certainly my opinion at that time." One journalist, probably Agnes's friend Thomas Ewing Dabney, wrote in 1947: "Mrs. Cleaveland, to all who know her, is a great lady, and her simplicity and kindliness are parts of that greatness." A family friend said of Agnes after her death: "She was thoughtful and kind and did many, many spontaneous loving actions. She was fearless in giving lifts to strangers in her car."

These friends might also have added that Agnes was highly intelligent, unpretentious, feisty, outgoing, community-minded, and nearly universally liked. Her sense of humor, which stayed with her almost to the end, was among her most endearing traits, as was her sense of adventure. Some might see her tendency to be outspoken as her greatest weakness; still others might consider this part of her charm.

From an early age, Agnes was a risk-taker, a trait that carried through into her eighties, when, for example, she chose to set up housekeeping at Cliffden despite her limited vision. She could have lived in more comfortable surroundings, yet like her sightless mother before her, Agnes wished to be independent. Later she had to accept help. Yet her determination at the end of her life to return to New Mexico and the place she called home was consistent with how she lived her entire life—she wanted to be in control.

Agnes Morley Cleaveland's greatest claim to fame is as author of *No Life for a Lady*, a book that continues to charm readers because of its realistic portrayal of ranch life, its good humor and cheerfulness, and its focus upon a strong-minded daredevil named Agnes. Though quite ill when she made her final trek to Howard Flat, she most likely would want to be remembered as the vibrant, free-spirited, strong-willed ranchwoman she once was—the one always "good enough to take along."

Sources

THE most important primary source for writing a biography of Agnes Morley Cleaveland is the Agnes Morley Cleaveland Papers, located in the Rio Grande Historical Collections [RGHC] in Branson Library at New Mexico State University in Las Cruces. The bulk of the papers falls within the years 1929–54. Included is correspondence with family members and with the general public. Especially valuable are the many fan letters Cleaveland received after publication of *No Life for a Lady*, her correspondence with Houghton Mifflin Company, and letters from Conrad Richter, Mary Roberts Coolidge, and Fred Lambert. The AMC Papers also contain a large number of her unpublished literary manuscripts, including drafts of essays about her brother Ray Morley, and many of her political essays. Very little in this collection pertains to Cleaveland's childhood or to her early married life; but newspapers clippings, AMC memoranda, and date books shed light on her club work and political activities in California.

Other collections in RGHC that have been important to this study are the Norman Cleaveland Papers, the Mary Cleaveland Wohlers Papers, and the Chase Ranch Records. A good source for Cleaveland family history is Norman Cleaveland's "Dredge Mining for Gold, Malaysian Tin, Diamonds, 1921–1966," an oral history conducted in 1994, three years before his death, by Eleanor Swent, Regional Oral History Office, the Bancroft Library, Berkeley, California. A copy of the transcript is available in RGHC. A few documents pertaining to AMC are found in the AMC Papers, Center for Southwest Research, University of New Mexico in Albuquerque. The Huntington Library in San

Marino, California, holds a small cache of Cleaveland family materials in the Norman Cleaveland Papers, as well as correspondence relating to W. R. Morley in the Chase Papers.

Correspondence between AMC and Lou Henry Hoover is located in the Herbert Hoover Presidential Library and Museum, West Branch, Iowa. Not only do the papers document their warm friendship but they also illuminate the political activism of both women. Also in the Hoover Presidential Library are a small number of letters between Herbert Hoover and AMC.

Other primary sources important to this study are the Dane Coolidge Papers at the Bancroft Library, the Conrad Richter Papers at Pennsylvania State University, deed books at the Socorro County Court House (New Mexico), and District Court records (San Miguel County), State Records Center and Archives, Santa Fe, New Mexico. I found invaluable the digitized newspaper files located through Ancestry.com and NewspaperArchive.com. These proved marvelous sources for following Cleaveland's career as clubwoman and political activist. Also important in tracing Cleaveland's movements in New Mexico were microfilmed copies of the following newspapers: *Las Vegas Daily Optic*, *Magdalena News*, *Silver City Independent*, *Socorro Bullion*, *Socorro Chieftain*, and the *Santa Fe New Mexican*. Personal reminiscences of Cleaveland or other members of the Morley and Cleaveland families were provided by Helen Cleaveland, Nancy Reed Miller, Mary Ann Montague, and Gary Tietjen.

The best source for AMC's youth and life on the ranch is, of course, *No Life for a Lady*. No one has seriously challenged the accuracy of her memoirs; many of the primary sources I've consulted confirm the authenticity of her stories. Nor has anyone written a scholarly account of Cleaveland's career as a writer, let alone of her career as clubwoman and political activist. Readers might find of interest two articles by Becky Jo Gesteland McShane, "In Pursuit of Regional and Cultural Identity, The Autobiographies of Agnes Morley Cleaveland and Fabiola Cabeza de Baca," in *Breaking Boundaries, New Perspectives on Women's*

Regional Writing, eds. Sherrie A. Inness and Diana Royer (Iowa City: University of Iowa Press, 1997), 180–96, and "Agnes Morley Cleaveland," in *Encyclopedia of Women in the American West*, eds. Gordon Morris Bakken and Brenda Farrington (Thousand Oaks, Calif.: Sage Publications, 2003), 52–55.

Still, the best essays on Cleaveland, which capture her personality shortly after she wrote *No Life for a Lady*, are Thomas Ewing Dabney's "Background for a Book," *New Mexico Magazine* 24 (July 1946): 24, 49–52, and "No Life for a Lady," *New Mexico Stockman* 11 (June 1946): 60, 77.

I relied on the following secondary sources for help in reconstructing the history of the Morley family prior to the death of William Raymond Morley: Keith L. Bryant, Jr., *History of the Atchison, Topeka & Santa Fe Railway* (Lincoln: University of Nebraska Press, 1974); David L. Caffey, *Frank Springer and New Mexico, From the Colfax County War to the Emergence of Modern Santa Fe* (College Station: Texas A&M University Press, 2006); Agnes Morley Cleaveland, *Satan's Paradise, From Lucien Maxwell to Fred Lambert* (Boston: Houghton Mifflin Co., 1952); Norman Cleaveland, *The Morleys, Young Upstarts on the Southwest Frontier* (Albuquerque: Calvin Horn Publisher, 1971); María E. Montoya, *Translating Property: The Maxwell Land Grant and the Conflict over Land in the American West, 1840–1900* (2002; reprint, Lawrence: University Press of Kansas, 2005); and Morris F. Taylor, *O. P. McMains and the Maxwell Land Grant Conflict* (Tucson: University of Arizona Press, 1979).

For the 1885 Indian scare in the Datils and the Camp on Datil Creek, I relied on Dan L. Thrapp, *The Conquest of Apacheria* (Norman: University of Oklahoma Press, 1967), and Letters Received, District of New Mexico, National Archives, Microfilm M 1088, rolls 58–60.

To understand educational opportunities for women at the end of the nineteenth century, I consulted Barbara Miller Solomon, *In the Company of Educated Women, A History of Women and Higher Education in America* (New Haven: Yale University Press, 1985). And for insight into campus life at the University of Michigan and Stanford University, I relied on Ruth Bordin,

Women at Michigan, The "Dangerous Experiment," 1870s to the Present (Ann Arbor: University of Michigan Press, 1999); Kent Sagendorph, *Michigan, The Story of the University* (New York: E. P. Dutton & Co., 1948); Orrin L. Elliott, *Stanford University, The First Twenty-Five Years* (Stanford University Press, 1937); and Margo Davis and Roxanne Nilan, *The Stanford Album, A Photographic History, 1885–1945* (Palo Alto, Calif.: Stanford University Press, 1989). Information about AMC's and Lou Henry's extracurricular activities at Stanford can be found in the *Stanford Quad* (student yearbook). Transcripts for Agnes Morley while at the University of Michigan were provided by the university; Agnes's Stanford transcripts were provided courtesy of Mary Ann Montague.

The following works provided insight into the western cattle industry during the late nineteenth and early twentieth centuries: Gerald Robert Baydo, "Cattle Ranching in Territorial New Mexico" (Ph.D. diss., University of New Mexico, 1970); Edward Everett Dale, *The Range Cattle Industry, Ranching on the Great Plains from 1865 to 1925* (Norman: University of Oklahoma Press, 1960); Maurice Frink et. al., *When Grass Was King, Contributions to the Western Range Cattle Industry Study* (Boulder: University of Colorado Press, 1956); Robert K. Mortensen, *In the Cause of Progress, A History of the New Mexico Cattle Growers' Association* (Albuquerque: New Mexico Stockman, 1983); David Remley, *Bell Ranch, Cattle Ranching in the Southwest, 1824–1947* (Albuquerque: University of New Mexico Press, 1993); and John T. Schlebecker, *Cattle Raising on the Plains, 1900–1961* (Lincoln: University of Nebraska Press, 1963).

See also Ruth W. Armstrong, *The Chases of Cimarron, Birth of the Cattle Industry in Cimarron Country, 1867–1900* (Albuquerque: New Mexico Stockman, 1981). Armstrong's work has to be used with caution. She does not cite specific sources, although the book is based on the Chase Ranch Records housed in RGHC. She also invents dialogue for people who appear in the text, including AMC and her mother, Ada Morley.

Especially helpful in understanding how Agnes Morley Cleaveland's experiences on the ranch compared to that of other women ranchers was Teresa Jordan's *Cowgirls, Women of the American West* (Garden City, New York: Anchor Press, 1982). But see also Nannie T. Alderson and Helena Huntington Smith, *A Bride Goes West* (1942; reprint, Lincoln: University of Nebraska Press, 1969); Eulalia Bourne, *Woman in Levi's* (Tucson: University of Arizona Press, 1967); and Mary Kidder Rak, *A Cowman's Wife* (1934; reprint, Austin: Texas State Historical Association, 1993). Two biographies of Caroline Lockhart, a contemporary of AMC who combined writing with ranching, are also noteworthy: Necah Stewart Furman, *Caroline Lockhart, Her Life and Legacy* (Seattle: University of Washington Press, 1994), and John Clayton, *The Cowboy Girl, The Life of Caroline Lockhart* (Lincoln: University of Nebraska Press, 2007).

For my understanding of Christian Science, I consulted Charles Braden, *Christian Science Today: Power, Policy, Practice* (Dallas: Southern Methodist University Press, 1958); Rennie B. Schoepflin, *Christian Science on Trial: Religious Healing in America* (Baltimore: Johns Hopkins University Press, 2003); and Rolf Swensen, "Pilgrims at the Golden Gate: Christian Scientists on the Pacific Coast, 1880–1915,"*Pacific Historical Review* 72 (May 2003): 229–63. AMC's own account of her membership in the church can be found in the AMC Papers at RGHC in an untitled manuscript, box 11, folder 11, and in a draft of a letter to Professor H. H. Powers, box 16, folder 13. For the Berkeley fire of 1923, I relied on newspaper accounts, AMC's observations, and Charles Wollenberg, *Berkeley, A City in History* (Berkeley: University of California Press, 2008).

AMC sustained a warm friendship with Eugene Manlove Rhodes. Unfortunately, very little of the correspondence between them has survived. But two works are especially helpful in following Rhodes's career: May Davison Rhodes, *The Hired Man on Horseback, My Story of Eugene Manlove Rhodes* (Boston: Houghton Mifflin Co., 1938), and W. H. Hutchinson, *A Bar Cross Man: The Life and Personal Writings of Eugene Manlove Rhodes* (Norman: University of Oklahoma Press, 1956).

For background material on Conrad Richter, I consulted the following: Conrad Richter, "The Sea of Grass," *New Mexico*

Magazine 43 (February 1965): 12–15; Conrad Richter, *Early Americana and Other Stories* (Boston: Gregg Press, 1978); Harvena Richter, *Writing to Survive: The Private Notebooks of Conrad Richter* (Albuquerque: University of New Mexico Press, 1988); and David R. Johnson, *Conrad Richter, A Writer's Life* (University Park: Pennsylvania State University Press, 2001).

Agnes Morley Cleaveland's involvement with the California Writers' Club is documented in the *Oakland Tribune* (California). Dave Sawle, president of the Berkeley Branch of the CWC, provided additional information via e-mail, 3/27/07 and 2/23/08.

The basic outline of Lou Henry Hoover's life is found in Nancy Beck Young's *Lou Henry Hoover, Activist First Lady* (Lawrence: University Press of Kansas, 2004). A full account of Hoover's involvement in Pro America has yet to be written, however. Biographical information about Herbert Hoover can be gleaned from his three-volume *Memoirs of Herbert Hoover* (New York: MacMillan Co., 1951–52). Of major importance to this biography of AMC were Joan Hoff Wilson's *Herbert Hoover, Forgotten Progressive* (Boston: Little, Brown, and Co., 1975), and Gary Dean Best's *Herbert Hoover, The Postpresidential Years, 1933–1964*, 2 vols. (Stanford, Calif.: Hoover Institution Press, 1983).

The best overview of women's clubs in America is Anne Firor Scott's *Natural Allies: Women's Associations in American History* (Urbana: University of Illinois Press, 1991). Scholarship dealing with women's clubs in California during the 1930s, when AMC was most active, however, is limited. Still, the following essays have been useful: Debra L. Hansen, "Clubs (Women's) in the West," and Clark Davis, "Clubwomen of Los Angeles," in *Encyclopedia of Women in the American West*, eds. Gordon Morris Bakken and Brenda Farrington (Thousand Oaks, Calif.: Sage Publications, 2003), 55–64, 73–79. For club activities of an earlier date, see Cameron Binkley, "'No Better Heritage Than Living Trees': Women's Clubs and the Early Conservation Movement in Humboldt County," *Western Historical Quarterly* 33 (summer 2002): 179–203; Mary S. Gibson, ed., *A Record of Twenty-five Years of the California Federation of Women's Clubs, 1900–1925* (Pasadena, Calif.: CFWC, 1927); Jerrod Harrison,

"After Suffrage: The California Federation of Women's Clubs and the 1913 State Legislature," located 9/8/08 at http://userwww.sfsu.edu/~epf/1997/harrison.html.

Works that helped place AMC's political activities in context include Anthony Arthur, *Radical Innocent: Upton Sinclair* (New York: Random House, 2006); June Melby Benowitz, *Days of Discontent, American Women and Right-Wing Politics, 1933–1945* (Dekalb: Northern Illinois University Press, 2002); James N. Gregory, *American Exodus, The Dust Bowl Migration and the Okie Culture in California* (New York: Oxford University Press, 1989); Daniel J. B. Mitchell, *Pensions, Politics, and the Elderly: Historic Social Movements and Their Lessons for Our Aging Society* (Armonk, N.Y.: M. E. Sharpe, 2000); Greg Mitchell, *The Campaign of the Century, Upton Sinclair's Race for Governor of California and The Birth of Media Politics* (New York: Random House, 1992); Catherine E. Rymph, *Republican Women: Feminism and Conservatism from Suffrage through the Rise of the New Right* (Chapel Hill: University of North Carolina Press, 2006); Kevin Starr, *Endangered Dreams: The Great Depression in California* (New York: Oxford University Press, 1996); and Susan Ware, *Holding Their Own: American Women in the 1930s* (Boston: Twayne Publishers, 1982).

I consulted the following to understand Earl Warren's importance to the California political scene in the 1930s: Leo Katcher, *Earl Warren, A Political Biography* (New York: McGraw-Hill Book Co., 1967), and Jim Newton, *Justice For All, Earl Warren and the Nation He Made* (New York: Riverhead Books, 2006). For Ada Morley's support of suffrage, see Joan M. Jensen, "'Disfranchisement is a Disgrace,' Women and Politics in New Mexico, 1900–1940," in *New Mexico Women; Intercultural Perspectives*, eds. Joan M. Jensen and Darlis A. Miller (Albuquerque: University of New Mexico Press, 1986), 301–31.

Kenneth O. May's story is told in Henry S. Tropp, "Kenneth O. May," *Isis* 70 (September 1979): 419–22, and Charles V. Jones, Philip Enros, and Henry S. Tropp, "Kenneth O. May, 1915–1977, His Early Life to 1946," *Historia Mathematica* 11 (November 1984): 358–79.

My account of books sent overseas during World War II is based on David G. Wittels, "What the G. I. Reads," *Saturday Evening Post*, June 23, 1945, pp. 11, 91–92, and John Y. Cole, ed., *Books in Action: The Armed Services Editions* (Washington, D.C.: Library of Congress, 1984). Information about Bertha Dutton's Girl Scout Mobile Archaeological Expeditions is in Jo Tice Bloom, "Dr. Bertha Dutton and Her Dirty Diggers," *La Crónica de Nuevo México* (October 2007): 1–3.

Note: For scholars and interested readers who may wish to check quotations and information found herein, a fully documented copy of my manuscript can be found in the Darlis A. Miller Papers, Hobson-Huntsinger University Archives, Branson Library, New Mexico State University, Las Cruces.

Index

References to illustrations are in italic type.